SEM ATLAS OF
CELLS AND TISSUES

TSUNEO FUJITA, M.D.

Professor of Anatomy,
Niigata University School of Medicine
Niigata, Japan

KEIICHI TANAKA, M.D.

Professor of Anatomy,
Tottori University School of Medicine
Yonago, Japan

JUNICHI TOKUNAGA, M.D.

Professor of Oral Bacteriology,
Faculty of Dentistry, Kagoshima University
Kagoshima, Japan

7/14/2000

IGAKU-SHOIN   Tokyo · New York

Published and distributed by

IGAKU-SHOIN Ltd.,
    5-24-3 Hongo, Bunkyo-ku, Tokyo
IGAKU-SHOIN Medical Publishers, Inc.,
    50 Rockefeller Plaza, New York, N. Y. 10020

ISBN: 0-89640-051-4
Library of Congress Catalog Card Number: 80-85298

Printed and bound in Japan

PREFACE

A decade has passed since a new visual world was introduced to the biomedical field by application of the scanning electron microscope (SEM). During this period a tremendously large variety of tissues, cells and other objects have been observed and a vast number of papers demonstrating their SEM images have been documented.

In the earlier half of this period (1968—1972), the *resolution* of the SEM was said to be 10—25 nm. This value falls two olders behind the resolution of the transmission electron microscope (TEM) (0.2nm) and several times behind the resolution of TEM images of replica membranes of biological specimens (5 nm). Practically, SEM observation was further hindered either by charging of the specimens or by the thick metal coating applied to prevent it. Nevertheless, the three-dimensional visualization of objects easily attained by the SEM turned out to be very useful for studies and valuable results rapidly accummulated in different research fields even in this period.

Our research group initiated studies with the SEM in 1967 and published an "Atlas of Scanning Electron Microscopy in Medicine" in 1970, which was a collection of SEM images covering not only tissues and cells of the body but also parasites, bacteria and fungi. This atlas thus served to demonstrate the broad applicability of the SEM to medical fields, but the micrographs presented in it were considerably low in quality judged from the present level. After this publication, the resolution of the SEM has been markedly elevated by different attempts, such as the invention of a SEM with a field-emission gun (Crewe, 1971), a SEM in a strong magnetic field (TEM-SEM) (Watanabe and Hoshino, 1976), and a low-loss image SEM (Broers *et al.*, 1975). Above all, the *field emission* type SEM which became commercially available from 1974 has enabled us to produce bright and smooth images of high resolution (1.5—2 nm at 25 kV, Nagatani and Okura, 1977) which roughly corresponds to that of replica images by TEM.

The main progress in specimen preparation was the spread of the *critical point drying* method on one hand, and on the other hand improvement of *fracture techniques* and *"staining"* methods giving the specimens electric conductivity.

This progress both in instruments and in specimen preparation techniques in the last five or six years has enabled us to produce SEM images whose quality is incomparably higher compared to the micrographs of five or six years ago.

We three authors of this volume have worked with the SEM in distant places of Japan but have continuously formed a hot team promoting mutual technical cooperation and information exchange. We now feel that it is timely to publish a collection of our micrographs prepared by the field emission microscopes and using the newest techniques of specimen preparation.

In this atlas the micrographs are arranged more systematically than in the previous publication, so that the readers might be able to review the entire microfabric of the body through the eye of the SEM. We believe that they contain much more information than previously concerning the structures of cellular and subcellular levels. Many of the images in this atlas may include hitherto

PREFACE

unknown features of cells waiting to be discovered by the readers. We intended to demonstrate the *interior structures of cells*, for the object of SEM observation today is not restricted to cell surfaces. Some *stereopairs* are included in order to fascilitate three-dimensional understanding of the structures concerned.

We authors would be most happy if this volume stimulates the interest of students and researchers into the marvelous structures of the microcosmos and encourages their study towards the elucidation of their functions and pathological changes. At the same time, we are confident that the micrographs in this atlas will represent the present-day level of SEM techniques and thus indicate a start-line for the progress of studies in the coming decade.

Many of the micrographs in this atlas were provided by courtesy of our colleagues, and we express our cordial thanks for their friendly help. We are especially indebted to Dr. M. Murakami, Professor of Anatomy at Kurume University Medical School; Dr. Y. Harada, Professor of Otolaryngology at Hiroshima University Medical School; Dr. T. Murakami, Associate Professor of Anatomy at Okayama University Medical School; Dr. Y. Uehara, Professor of Anatomy at Ehime University Medical School; Dr. T. Masutani, Department of Anatomy at Fukuoka University Medical School and Dr. M. Edanaga at the Research Laboratories of Yoshitomi Pharmaceutical Industries, Ltd.

Sincere thanks are due to our technical staff, especially to Mr. Y. Kashima and Mr. H. Osatake at Tottori University School of Medicine, and to Mr. K. Adachi at Niigata University School of Medicine for their skillfull and enthusiastic cooperation in scanning; also, to Miss Sachiko Ozawa for her patient darkroom work and to Miss Tomoko Kawakami for her help in typing and editing the manuscript.

Finally, we express our cordial thanks to the staff of the publisher, Igaku Shoin, especially to Mr. H. Okada, Mr. M. Sakamoto and Mr. M. Takeda, for their kind cooperation.

CONTENTS

METHODS

A large variety of techniques are known for preparing SEM specimens of different kinds and it would need too many pages to cover them. This chapter will therefore concentrate on introducing the methods used in the work for this atlas, mentioning only popular, related techniques and some important literature.

Washing and Perfusion

The specimen surface, whether natural or artificial, to be observed under the SEM must be free of mucus or protein-rich body fluids or other contamination. The use of some enzymes like papain (Ferguson and Heap, 1970) and N-acetyl-L-cystine (Boyde and Wood, 1969) as well as supersonic vibration has been recommended for elimination of mucous coverage. Generally, however, a careful, intensive rinse with a jet of saline gives a satisfactory effect. Rinsing should be repeated after aldehyde fixation also.

In most tissues, especially those highly vascularized organs like the spleen, liver and kidney, *vascular perfusion* with a large amount of warmed Ringer solution is most recommended. This procedure not only eliminates blood and exposes the luminal surface of blood vessels, but also cleans the surface of the tissue elements because the saline partly oozes out from the vessels to run through the extravascular spaces. Perfusion with saline usually is followed by that with a fixative (*vide infra*).

Fixation

In the early stage of SEM application to biological materials the fixation method was considered not very important to the quality of images, but along with the elevated resolution of the instruments, it became eagerly questioned as to what kind of fixation should be selected in combination with what kind of drying techniques for a given tissue or cell (Arenberg *et al.*, 1970; Shimamura and Tokunaga, 1970; Tokunaga, 1970; Arnold *et al.*, 1971; Boyde and Vesley, 1972; Litke and Low, 1977).

Glutaraldehyde and *formaldehyde* (solution of paraformaldehyde) are usually used as tissue fixatives for SEM. As is well known, the former surpasses the latter in fixation power and ultrastructure preservation, while the latter is better in penetration rate and in preservation of some enzyme activities. To compensate for the weakpoints of either aldehyde, mixtures of glutaraldehyde and formaldehyde (e.g., diluted Karnovsky's fixative; cf. Litke and Low, 1977) are useful for SEM materials.

Acrolein has an incomparably rapid penetration rate and strong fixative effect, though it causes moderate contraction of the tissue. This aldehyde is useful in immersion fixation of large or compact tissues. Previous TEM studies suggested that a mixture of acrolein and glutaraldehyde provides better preservation of cytoplasmic structures than either aldehyde alone (Sandborn, 1966; Hayat, 1968).

The *osmotic pressure* of the fixatives affects the shape of the cells considerably. For TEM specimens moderately hypertonic fixatives (400—500 milliosmols) are favored (Fahimi and Drochman, 1965). In SEM also, slightly hypertonic fixatives are recommended for preservation of natural cell shapes as indicated by the examination of Hattori *et al.* (1975) using erythrocytes. Litke and Low (1977) conceive that for adequate SEM fixation, "the fixative vehicle should be approximately isotonic for tissue, with aldehyde (2% or less) added to vehicle".

For the above reasons, the fixatives used in the present work were:
1. Glutaraldehyde, 1 or 2%, in 0.1M phosphate buffer, pH 7.3.
2. Glutaraldehyde, 1%, and paraformaldehyde, 1% in 0.1M phosphate buffer, pH 7.3.
3. Acrolein, 2.5%, in 0.1M phosphate buffer, pH 7.3.
4. Acrolein, 1.5%, glutaraldehyde, 3% and paraformaldehyde, 1.5% in 0.2M cacodilate buffer, pH 7.2.

Fixation of tissue may be made by *immersion* either after it is taken out of the body (*in vitro*) or as it is within the body (*in situ*). However, perfusion fixation *in situ* gives in most cases the best result, as it ensures rapid and even penetration of fixatives avoiding anoxic reactions and mechanical deformation of the tissue.

Perfusion fixation is made usually from the heart, aorta or an artery supplying the tissue to be fixed. The portal vein, for the liver, and certain other vessels may be used. Perfusion with saline (*vide supra*) is in most cases recommended beforehand. The tissue thus fixed is later excised and cut into small blocks, which are then immersed in the same fixative.

Postfixation with OsO_4, or postosmication, has been introduced to the SEM methodology firstly to harden the tissue and prevent its deformation by surface tension during air drying and secondly to give the specimen electric (and thermal) conductivity and increase secondary emission from the specimen. Recently, however, osmication for the first purpose is not needed, as the critical point method has resolved the problem of specimen damage during drying. Postosmication for the second purpose has increased in necessity and has been developed into techniques which are now collectively called conductive staining.

Conductive Staining

Special techniques have been known to give electric conductivity to biological specimens in order to prevent charging during scanning. These techniques aim at the same time to impregnate the specimens with a material amply emitting secondary emissions. Specimens treated with these techniques can be observed in the SEM without unfavorable charging effects using even higher accelerating voltage like 25 kV. Such specimens are especially suited for microdissection within the SEM (Murakami *et al.*, 1977). For routine scanning, however, a thin metal coating is applied.

The *tannic acid-osmium method* developed by Murakami (1973, 1974) seems to be one of the best along this line and was used for most of the micrographs in this atlas. The procedures (partly simplified) and principles of the method are as follows:

Aldehyde-fixed tissues are put for 5—16 hrs in an aqueous solution composed of 2% glycine, 2% sodium glutamate and 2% sucrose (pH 6.2). The amino acids and sugar are thus connected with the free aldehyde groups of the fixative remaining in the tissue. After being rinsed in water, the tissues are immersed for 6—24 hrs in a 2% aqueous solution of tannic acid (pH 4.0). During this process, an ample amount of tannic acid combines with the proteins and polysaccharides of the tissue and with the amino acids and sugar previously implanted. After being rinsed in water for 1 hr the tissues are immersed for 6—12 hrs in a 2% aqueous solution of osmium tetroxide. The tissue is thus heavily impregnated with osmium tetroxide, which combines with tissue proteins and lipids on one hand and with tannic acid forming chelates on the other hand.

If a quick procedure is desired, the first step using the solution of amino acids and sugar may be simply omitted. Immersion in tannic acid and in osmium may be repeated if the tissue is not sufficiently impregnated after one course of the procedures.

Some *other conductive stainings*, especially the method by Kelley *et al.* (1973) using thiocarbohydrazide as a bridge between the tissue and osmium, also give good results.

Methods using salt solutions of metals other than osmium may be effective as conductive stainings. Saturated solution of uranium acetate is quite useful if one leaves small pieces of aldehyde-fixed tissues in it overnight. After a simple rinse, the slightly brown-tinged tissues can be processed to dehydration and fracture procedures (Tokunaga, unpublished).

The conductive stainings, by either method mentioned above, ensure good conductivity and ample secondary emission of the specimens, and represent the indispensable procedure for high resolution SEM even when metal coating is given to the specimens.

Cracking

The natural surface of a specimen is only a small part of the object of SEM observation. By cutting open the specimens, the world of SEM images has been tremendously expanded. We can see the surface structures of cells hidden deeply in the tissue and, moreover, advanced fracture techniques have included the interior structures of cells within the range of high resolution SEM.

The simplest and yet very useful cracking method is to break the tissue by hands or forceps after aldehyde fixation or after osmication; occasionally dried specimens may be cracked even more effectively. This method is useful for observation of the cellular construction of organs and some surface structures of cells as the fracture usually occurs along cell boundaries. With this method, however, the fracture surface as a whole is quite rough and often contaminated with broken material. Moreover, structures may be moved from their original site; fibrous matter may be pulled out.

In order to create a clean and precise fracture surface, Tanaka (1972) originated a *"frozen resin cracking method"*. Fixed tissue pieces are, through ethanol and propylene oxide, dehydrated and placed in gelatine capsules filled with Cemedine 1500 (epoxy resin, Cemedine Co., Tokyo) without adding any catalysts. This matter is cooled for 1—2 hrs at −30°C and cracked with a chisel and hammer. A smooth, glittering surface is thus produced.

Tanaka *et al.* (1974) also published a *"styrene resin cracking method"* in which the ethanol dehydrated specimens were immersed for 30 min in a 1:1 mixture of styrene and ethanol and then left in cold (4°C) styrene overnight. The specimens are embedded in gelatin capsules filled with styrene monomer containing 2—3% benzoylperoxide (catalyst). Polymerization is caused by heating at 60°C for 24 hrs and the specimens are cracked at room temperature.

In both frozen and polymerized resin methods, a small guillotine-like apparatus devised by Tanaka *et al.* (1974) may be useful for obtaining a fracture of the (approximately) desired direction and site. In both methods, the fractured specimens are cleaned of resin in propylene oxide at room temperature, and, after immersion in amylacetate, the specimens are critical point dried. These methods, used partly in the present work, produce clean fractures of cells and their organelles (Tanaka *et al.*, 1976).

As a simpler method, Tanaka's group introduced a *"frozen liquid cracking method"* (Hamano *et al.*, 1973). Fixed tissues were immersed in ethanol, quench-frozen in liquid nitrogen and cracked. The tissues then were processed to critical point drying. A closely similar method was proposed also by Humphreys *et al.* (1974).

Tokunaga *et al.* (1974) extended the study along this line and found some technical modifications to be useful for high resolution SEM of tissues and cells. These techniques have been applied to the majority of the specimens used in the present work and will be introduced here:

Tissues, after being fixed and conductive-stained, are dehydrated and immersed in ethanol or amylacetate, frozen in liquid nitrogen or in Freon 22 cooled by liquid nitrogen and are cracked with a small chisel. Clean and fairly smooth fracture surfaces are obtained by this method, but sometimes they are too smooth and little information may be obtained from the fractured portion itself. If the tissues are, without dehydration, immersed in a 40% aqueous solution of dimethyl sulfoxide (DMSO, which has been known as an agent which prevents the occurrence of ice crystal damage) and freeze-cracked as mentioned above, the fracture occurs mainly along the cell surfaces and intracellular membranes. More plastic information on cellular and subcellular structures may be obtained by this method.

Drying

Water is the most important constituent of living cells both physically and chemically, and it required much endeavor to remove water from the biological specimens without unfavorable deformation.

To lessen the enormous surface tension which during air drying of wet specimens causes severe deformation, drying through a *volatile medium* like acetone, ethanol, ether or amylacetate was widely used in the early days (Barber and Boyde, 1968; Fujita *et al.*, 1968). Because of its simple procedure, this method may still be useful for certain purposes like quick examination of clinical specimens. *Freeze drying* has been applied to SEM specimens (Blümcke and Morgenroth, 1967; Boyde and Barber, 1969) but is not widely used now because of its time consuming procedure (several days for tiny tissue pieces) and the occasional occurrence of ice crystal damage.

Critical point drying is now most favored in SEM studies as the best method, both practically and theoretically, to avoid the effects of surface tension during the drying of specimens (Boyde and Wood, 1969; Horridge and Tamm, 1969). As water has a critical temperature (Tc) of 374°C and a critical pressure (Pc) of 218 atm, it is not suited either technically or biologically. Water, thus must be substituted by a substance with a lower Tc and Pc. Use has been made of CO_2 (introduced by Anderson, 1951 for drying blood and sperm cells in their natural form), N_2O (Koller and Bernhard, 1964), Freon 13 ($CCl F_3$; Cohen *et al.*, 1968), Freon 22 ($CCl F_2$) and SO_2 (Turner and Green, 1973). Because of its low price and non-toxicity, CO_2 is most widely used for SEM specimens. Its Tc is 31.3°C and the Pc is 72.9 atm; the latter may be claimed a little too high for the sake of safety in daily laboratory use. Freon 13 has a Tc of 28.9°C and a Pc of 38.2 atm and thus, in this respect, is easier to use than CO_2. However, the relatively high price of freon hinders its popular use.

The CO_2 critical point method consists of the following procedures: fixed (and conductive-stained) tissues are dehydrated by graded concentrations of ethanol, immersed in amylacetate and transferred into a small pressure chamber. Liquid CO_2 is introduced into the chamber and the amylacetate is replaced by the CO_2. The last procedure may be replaced by a simple method proposed by Tanaka and Iino (1974) using dry ice instead of high-pressure CO_2.

The chamber is then warmed to above the critical point. If the chamber has an observation window, one may see, around the critical point, specimens surrounded by waving filaments of air which represent the intermediate stage between liquid and gas. The specimen medium is transformed from liquid to gas without forming a boundary, that is, without causing surface tension. The gas is then slowly let out of the chamber and the dried specimens are removed.

Coating

Some biological specimens can be observed by SEM *without* metal coating by the use of low accelerating voltage (1—5 kV). This can be done even with higher voltage if the specimens are scanned more rapidly by a TV-scanning system. As mentioned above, conductive-stained specimens can be scanned without unfavorable charging under high accelerating voltage such as 25 kV.

Yet it is usual that the specimens are "conductive-coated" with metals to ensure electric and thermal conductivity of the specimen and to make a thin layer from which the secondary emission is produced.

In the early stage of these studies, a double coating with carbon and gold was recommended (Barber and Boyde, 1968), and the simple use of gold also prevailed because of the easy and quick procedure. However, gold turned out to be unsuited to high resolution SEM because it, when evaporated upon the specimen, forms unevenly distributed particles and, furthermore, these particles tended to accumulate into coarse masses. In this respect gold-palladium (Au:Pd = 60:40) is most recommended and also used in the present work because this alloy forms a fine-granular and homogeneous coating membrane which is relatively stable under the electron beam. Platinum-palladium (Pt:Pd = 90:10) has the same advantages though somewhat less easy to use (Nagatani and Saito, 1974).

Vacuum evaporation is used for the coating of specimens with the metals mentioned above. In order to attain a coating layer as even and continuous as possible, the specimens are rotated and tilted in two mutually perpendicular directions during evaporation of the metal (Barber and Boyde, 1968).

Ion sputtering recently has become popular as a simple and effective coating technique (Echlin, 1975, DeNee and Walker, 1975). In a low vacuum (1—5×10^{-2} Torr) containing an inactive gas (nitrogen, argon, etc.), the specimens are placed on the anode and the metal to be sputtered forms the cathode. Under electric voltage of 1.0—1.2 kV (5—10 mA) the gas is ionized; positive ions hit the cathode and sputter metal atoms, which are deposited on the specimens. Many kinds of metal can be sputtered by this method and, as the deposition rate is slow, one can control the thickness of the metal coat more easily than with evaporation 0.1nm/sec), one can control the thickness of the metal coat more easily than with evaporation techniques (Echlin, 1975). The metal attached to the specimen by sputtering generally seems to be finer grained than that by evaporation. However, delicate portions of biological specimens may sometimes show damage from metal ion bombardment, like the occurrence of tiny pits. Because of this danger, most of the specimens in the present work have been coated by the evaporation method.

The *thickness of the metal coat* should vary according to the nature of the specimen and the purpose of observation. Although a metal coat of 10—30 nm was favored in early studies (Barber and Boyde, 1968), today's high resolution SEM, whose resolution approximates 2 nm, demands a coating of the thickness of this order (Tokunaga *et al.*, 1976). As stressed repeatedly, special measures such as conductive staining of the specimen are thus inevitably required.

Ion-Etching

Ion-etching is a useful method to eliminate soft materials and contamination covering the specimen, or, more actively, to "dig out" certain interior structures resistant to etching. A few figures (Fig. 1-6; Plate I-12A; Plate I-13B) in this atlas showing intracellular structures have been produced by the ion-etching technique.

Ion-etching is achieved by the ejection of atoms from the material surface bombarded by ions. The rate of atom ejection depends: (1) upon the mass and velocity of the ions, (2) upon the mass of the atoms to be ejected, and (3) upon the energy combining the atoms to the material surface.

In order to reveal the intrinsic structures in biological specimens by ion-etching, it is essential to avoid the unidirectional cone formation (Lewis *et al.*, 1968; Hodges *et al.*, 1972; Fulker *et al.*, 1973) either by tilting the specimens or by using ions of various energies and directions. For this purpose the following method is used by our research group (Fujita *et al.*, 1974):

In a simple glow discharge apparatus, an alternating current discharge (50 Hz) is made between two aluminum plates under very low vaccum (0.05—0.5 Torr). The electrode voltage is about 700 V and the current about 2 mA. Ions, mainly of nitrogen and oxygen, thus produced are applied for 1—2 hrs to fixed and dried specimens placed on one of the electrode plates. With this method we can eliminate the unidirectional effect of etching without tilting the specimens. The etched specimens are then coated with metals.

Selected New Techniques

1. Tanaka's method for cell maceration

Quite recently a very useful and simple technique was introduced by Tanaka and Naguro (1981). In this method, which Tanaka calls "Os-DMSO-Os method", soft part of cytoplasm or cytoplamic matrix of fractured cells is removed by the macerating effect of OsO_4, and membranous and solid portions of cell organelles are exposed as shown in Plate I-10. The procedure is as follows:

1. The tissue is fixed in 1% OsO_4 in 1/15 M phosphate buffer (pH 7.4), for 1—2 hrs.
2. The tissue is successively dipped in 25% and 50% DMSO aqueous solution and freeze-fractured.
3. The tissue is kept in 0.1% OsO_4 for 24—72 hrs.
4. Rinsed and critical point dried.

2. Evan's method for cell denudation

Visualization of cellular surfaces covered by connective tissue or basement membrane has been enabled by Evan and his associates (1976). They succeeded in elimination of collagen and related structures by treating the fixed specimens with warmed HCl and then with collagenase. The procedures are as follows:

1. Specimens are rinsed several times with phosphate buffer and the fixative in them is removed.
2. The specimens are placed in 8 N HCl for 50—70 min at 60°C. One must carefully find appropriate timepoint to stop this maceration for every given tissue. Treatment longer than the critical time causes dissolution of the tissue.
3. Microdissection, if necessary, is performed, and the specimens are placed in phosphate buffered collagenase, pH 6.8, for 3—8 hrs at 37°C.

4. After several rinse the specimens are processed to critical point drying.

The most important point in the Evan's technique is the use of warm 8 N HCl. With this procedure only, without the use of collagenase, cells may be beautifully denuded from basement membrane as shown in the pictures of Plate XI-3.

Treatment of Free Cells

Preparation methods for free cells such as blood cells and spermatozoa will be introduced here only briefly because only a few micrographs concerning these objects are included in this atlas.

For fixation of these cells 1% glutaraldehyde in 0.1 M phosphate buffer is generally recommended because this slightly hypertonic fixative satisfactorily preserves the natural shape of the cells, as far as the experiments using erythrocytes indicate (Hattori *et al.*, 1969).

The cells may be washed, before fixation, with saline but more simply, one may drop a small amount of blood, semen or a given fluid into a large amount of aldehyde fixative being shaken. The cells are thus washed and fixed simultaneously, while the blood and sperm plasma is diffused extensively and eliminated by centrifugation (Tokunaga *et al.*, 1969; Fujita *et al.*, 1970).

After postosmication (conductive staining) and dehydration the free cells are suspended in amylacetate, dropped on a small piece of glass slide and transferred into the pressure chamber for critical point drying. A slight amount of amylacetate should remain on the specimen when the liquid CO_2 is introduced, but if the former is too much in amount the cells will float away with the liquid CO_2.

To prevent this disadvantage *light centrifugation* of a cell suspension may be done either before or after fixation onto a piece of glass slide placed in the bottom of centrifuge tube. The slide to which the cells are attached can then be dehydrated and critical point dried quickly and safely.

Another more advanced method is to ensure attachment of the cells by coating the glass surface with a *positively charged substance;* cell surfaces are negatively charged. Usually glass slides are coated with poly-L-lysine (Tsutsui *et al.*,1976). The slides are placed for 5 min in 0.1% solution (0.1M phosphate buffer, pH 7.4) of this polycationic substance and simply rinsed with the buffer. Fresh or fixed cells to be observed are placed in contact with the slides for 5 min, and they become amply attached to the latter. The slides, after rinsing, are processed to critical point drying.

The use of *filters* of different kinds to gather and sustain free cells has been tried by many researchers. Recently polycarbonate membranes with pores of given diameter (Nuclepore, General Electric Co.) have become available. Because of their insolubility in the solvents used for drying and the smooth surface with uniform pores when viewed under the SEM, they are proving to be good sustainers of free cells. One may simply let a cell suspensions pass the membrane with pores of appropriate size, then process the membrane to critical point drying.

Metal coating of free cells usually must be made more carefully than that of tissue blocks, as electric conductivity of cells lightly attached to the sustainer is difficult to insure.

Vascular Casts

Attempts have been made to observe by SEM the casts of canalicular structures (Nowell *et al.*, 1970; Tyler *et al.*, 1970; Murakami, 1971). Especially good results have been produced by Murakami (1971, 1973) in his *methyl methacrylate replicas* of blood vessels. Our knowledge of microcirculation in different organs is markedly advancing by this method. The procedure for vascular casting is as follows:

Methyl methacrylate ester monomer (100 ml) to which is added 1—1.5 ml 2,4-dichlorobenzoyl peroxide (catalyst) is warmed to 60—65°C. As polymerization is initiated, it will spontaneously reach 85—95°C in a few minutes. This half-polymerized resin is then rapidly cooled to a temperature lower than 30°C. One and a half to 2 g benzoyl peroxide is then added and after it is gently stirred for 5—10 min it should become suitably viscous (slightly less viscous than glycerin) at which time 1.5 ml dimethyl aniline (accelerator) is added to the mixture. After the animal or desired organ is arterially perfused with Ringer solutions of gradually ascending temperatures (37—50°C), the mixture is injected with a moderate amount of pressure. The animal or organ is immersed in a hot water bath (60—70°C) and placed in an oven of the same temperature for 24 hrs in order to polymerize the resin. The animal tissue is then macerated with 20% NaOH and the resin cast is washed thoroughly.

Recently a methacrylate medium prepared for casting of blood capillaries has become commercially available (Mercox; Japan Vilene Co. Ltd.).

The methacrylate casts may be coated with metals, but after being placed in osmium gas for a few hours, they obtain excellent conductivity and give beautiful images even without any coating (Murakami, 1973).

Scanning

Most of the micrographs in this volume were taken under accelerating voltage of 10 kV but some of them with 25 kV. Field emission type scanning electron microscopes (Hitachi HFS-2 and HFS-2S) were used.

Stereopairs were photographed with a tilting angle of 10°.

References

Arenberg, I. K., W. F. Marovitz and A.P. Mackenzie: Preparative techniques for the study of soft biological tissues in the scanning electron microscope: A comparison of air drying, low temperature evaporation, and freeze drying. Proc. 3rd Annual Stereo Scan Colloquium, 1970. (p.121—142).

Arnold, J. D., A. E. Berger and O. L. Allison: Some problems of fixation of selected biological samples for SEM examination. In: (ed. by) O. Johari and I. Corvin: Scanning Electron Microscopy/1971. IIT Research Institute, Chicago, 1971. (p. 249—256).

Barber, V. C. and A. Boyde : Scanning electron microscopic studies of cilia. Z. Zellforsch. 84: 269—284 (1968).

Blümcke, S., K. Morgenroth, Jr.: The stereo ultrastructure of the external and internal surface of the cornea. J. Ultrastr. Res. 18: 502—518 (1967).

Boyde, A. and V. C. Barber : Freeze-drying methods for the scanning electron-microscopical study of the protozoon *Spirostomum ambiguum* and the statocyst of the cephalopod mollusc *Loligo vulgaris*. J. Cell Sci. 4 : 223—239 (1969).

Boyde, A. and P. Vesely: Comparison of fixation and drying procedures for preparation of some cultured cell lines for examination in the SEM. In: (ed. by) O. Johari and I. Corvin: Scanning Electron Microscopy/1972. IIT Research Institute, Chicago, 1972. (p.265—272).

Boyde, A. and C. Wood: Preparation of animal tissues for

surface scanning electron microscopy. *J. Microscopy* 90: 221—249 (1969).

Broers, A. N., B. J. Panessa and J. F. Gennaro, Jr.: High resolution SEM of biological specimens. In: (ed. by) O. Johari and I. Corvin: Scanning Electron Microscopy/1975. IIT Research Institute, Chicago, 1975. (p. 233—242).

Cohen, A. L., D. P. Marlow and G. E. Garner: A rapid critical point method using fluorocarbons ("Freons") as intermediate and transitional fluids. *J. Microscopie* 7: 331—342 (1968).

Crewe, A. V.: High resolution scanning microscopy of biological specimens. *Phil. Trans. Roy. Soc. Lond.* 261: 61—70 (1971).

DeNee, P. B. and E. R. Walker: Specimen coating technique for the SEM — A comparative study. In: (ed. by) O. Johari and I. Corvin: Scanning Electron Microscopy/1975. IIT Research Institute, Chicago, 1975. (p. 225—232).

Echlin, P.: Sputter coating techniques for scanning electron microscopy. In: (ed. by) O. Johari and I. Corvin: Scanning Electron Microscopy/1975. IIT Research Institute, Chicago, 1975. (p.217—224).

Evan, A. P., W. G. Dail, D. Dammrose and C. Palmer: Scanning electron microscopy of cell surfaces following removal of extracellular material. *Anat. Rec.* 185: 433—446 (1976).

Evan, A., W. G. Dail, D. Dammrose and C. Palmer: Scanning electron microscopy of tissues following removal of basement membrane and collagen. In: (ed. by) O. Johari and R. P. Becker: Scanning Electron Microscopy/1976/II. IIT Research Institute, Chicago, 1976 (p. 204—208).

Fahimi, H. D. and P. Drochmans: Essais de standardisation de la fixation au glutaraldéhyde. I. Purification et détermination de la concentration du glutaraldéhyde. II. Influence des concentrations en aldéhyde et de l'osmolalité. *J. Microscopie* 4: 725—748 (1965).

Ferguson, D. R. and P. F. Heap: The morphology of the toad urinary bladder: A stereoscopic and transmission electron microscopical study. *Z. Zellforsch.* 109: 297—305 (1970).

Fujita, T., H. Inoue and T. Kodama: Scanning electron microscopy of the normal and rheumatoid synovial membranes. *Arch. histol. jap.* 29: 511—522 (1968).

Fujita, T., M. Miyoshi and J. Tokunaga: Scanning and transmission electron microscopy of human ejaculate spermatozoa with special reference to their abnormal forms. *Z. Zellforsch.* 105: 483—497 (1970).

Fujita, T., T. Nagatani and A. Hattori: A simple method of ion-etching for biological materials. An application to blood cells and spermatozoa. *Arch. histol. jap.* 36: 195—204 (1974).

Fulker, M. J., L. Holland and R. E. Hurley: Ion etching of organic materials. In: (ed. by) O. Johari and I. Corvin: Scanning Electron Microscopy/1973. IIT Research Institute, Chicago, 1973. (p.379—386).

Hamano, M., T. Otaka, T. Nagatani and K. Tanaka; A frozen liquid cracking method for high resolution scanning electron microscopy (Abstr.). *J. Electron Microsc.* 22: 298 (1973).

Hattori, A., S. Ito, A. Sugawara and M. Matsuoka: Studies on fixation and drying of blood cells for scanning electron microscopic observation. (Japanese text with English abstract.) *Acta haematol. jap.* 38: 86—95 (1975).

Hattori, A., J. Tokunaga, T. Fujita and M. Matsuoka: Scanning electron microscopic observations on human blood platelets and their alterations induced by thrombin. *Arch. histol. jap.* 31: 37—54 (1969).

Hayat, M. A.: Triple fixation for electron microscopy. Proc. 26th Ann. Meet. Electron. Microsc. Soc. Amer., Claitor's Publishing Division, Baton Rouge, 1968 (p.90).

Hodges, G. M., M. D. Muir, C. Sella and A. J. P. Carteaud: The effect of radio-frequency sputter ion etching and ionbeam etching on biological material. A scanning electron microscope study. *J. Microscopy* 95: 445—451 (1972).

Horridge, G. A. and S. L. Tamm: Critical point drying for scanning electron microscopic study of ciliary motion. *Science* 163: 817—818 (1969).

Humphreys, W. J., B. D. Spurlock and J. S. Johnson: Critical point drying of ethanol-infiltrated, cryofractured biological specimens for scanning electron microscopy. In: (ed. by) O. Johari and I. Corvin: Scanning Electron Microscopy/1974. IIT Research Institute, Chicago, 1974. (p. 275—282).

Kelley, R. O., R. A. F. Dekker and J. G. Bluemink: Ligand-mediated osmium binding: Its application in coating biological specimens for scanning electron microscopy. *J. Ultrastr. Res.* 45: 254-258 (1973).

Koller, T. and W. Bernhard: Séchange de tissus au protoxyde d'agote (N$_2$O) et coupe ultrafine sans matière d'inclusion. *J. Microscopie* 3: 589—606 (1964).

Lewis, S. M., J. S. Osborn and P. R. Stuart: Demonstration of an internal structure within red blood cell by ion etching and scanning electron microscopy. *Nature* 220:614—616 (1968).

Litke, L. L. and F. N. Low: Fixative tonicity for scanning electron microscopy of delicate chick embryos (1). *Amer. J. Anat.* 148: 121—127 (1977).

Murakami, T.: Application of the scanning electron microscope to the study of the fine distribution of the blood vessels. *Arch. histol. jap.* 32: 445—454 (1971).

Murakami, T.: A metal impregnation method of biological specimens for scanning electron microscopy. *Arch. histol. jap.* 35: 323—326 (1973).

Murakami, T.: A revised tannin-osmium method for non-coated scanning electron microscope specimens. *Arch. histol. jap.* 36: 189—193 (1974).

Murakami, T., K. Yamamoto, T. Itoshima and S. Irino: Modified tannin-osmium conductive staining method for non-coated scanning electron microscope specimens. Its application to microdissection scanning electron microscopy of the spleen. *Arch. histol. jap.* 40: 35—40 (1977).

Nagatani, T and A. Okura: Enhanced secondary electron detection at small working distance in the field emission SEM. In: (ed. by) O. Johari: Scanning Electron Microscopy/1977/I. IIT Research Institute, Chicago 1977 (p. 695—702).

Nagatani, T. and M. Saito: Structure analysis of evaporated films by means of TEM and SEM. In: (ed. by) O. Johari and I. Corvin: Scanning Electron Microscopy/1974. IIT Research Institute, Chicago, 1974. (P. 51—58).

Nowell, J. A., J. Pangborn and W. S. Tyler: Scanning electron microscopy of the avian lung. In: (ed. by) O. Johari and I. Corvin: Scanning Electron Microscopy/1970. IIT Research Institute, Chicago, 1970. (p. 249—256).

Sandborn, E.: Electron microscopy of the neuron membrane systems and filaments. *Can. J. Physiol. Pharmacol.* 44: 329 (1966).

Shimamura, A. and J. Tokunaga: Scanning electron microscopy of sensory (fungiform) papillae in the frog tongue. In: (ed. by) O. Johari and I. Corvin: Scanning Electron Microscopy/1970. IIT Research Institute, Chicago, 1970. (p.225—232).

Tanaka, K.: Freezed resin cracking method for scanning electron microscopy of biological materials. *Naturwiss.* 2: 77 (1972).

Tanaka, K. and A. Iino: Critical point drying method using Dry Ice. *Stain Technol.* 49: 203—206 (1974).

Tanaka, K. and T. Naguro: High resolution scanning electron microscopy of cell organelles by a new specimen preparation method. *Biomed. Res.* 2 (Suppl.): 63—70 (1981).

Tanaka, K., A. Iino and T. Naguro: Styrene resin cracking method for observing biological materials by scanning electron microscopy. *J. Electron Microsc.* 23: 313—315 (1974).

Tanaka, K., A. Iino and T. Naguro: Scanning electron microscopic observation on intracellular structures of ion-etched materials. *Arch. histol. jap.* **39**: 165—175 (1976).

Tanaka, T., N. Kosaka, T. Takiguchi, T. Aoki and S. Takahara : Observation on the cochlea with SEM. In: (ed. by) O. Johari and I. Corvin: Scanning Electron Microscopy/1973. IIT Research Institute, Chicago, 1973. (p. 427—434).

Tokunaga, J.: The preparation for specimens of soft tissues and cells for scanning electron microscope (Abst.) *J. Electron microsc.* **19**: 108 (1970).

Tokunaga, J., M. Edanaga, T. Fujita and K. Adachi: Freeze cracking of scanning electron microscope specimens. A study of the kidney and spleen. *Arch. histol. jap.* **37**: 165—182 (1974).

Tokunaga, J., T. Fujita and A. Hattori: Scanning electron microscopy of normal and pathological human erythrocytes. *Arch. histol. jap.* **31**: 21—35 (1969).

Tokunaga, J., T. Fujita, A. Hattori and J. Müller: Scanning electron microscopic observation of immunoreactions on the cell surface: Analysis of *Candida albicans* cell wall antigen by the immunoferritin method. In: (ed. by) O. Johari and I. Corvin: Scanning Electron Microscopy/1976 I. IIT Research Institute, Chicago, 1976. (p. 301—310).

Tsutsui, K., H. Kumon, H. Ichikawa and J. Tawara: Preparative method for suspended biological materials for SEM by using of polycationic substance layer. *J. Electron Microsc.* **25**: 163—168 (1976).

Turner, R. H. and C. D. Green: Preparation of biological material for scanning electron microscopy by critical point drying from water miscible solvents. *J. Microscopy* **97**: 357—363 (1973).

Tyler, W. S., J. A. Nowell and J. Pangborn: Techniques for scanning electron microscopy of tissues and replicas. Proc. 7me Congr. Int. Microsc. Electron. Grenoble. 1970. (p. 477—478).

Watabe, T. and T. Hoshino: Observation of individual ferritin particles by means of scanning electron microscope. *J. Electron Microsc.* **25**: 31—33 (1976).

CHAPTER 1

CELL

Cell Surface Structure

Our knowledge of cell surface structure has rapidly been enriched by the advance of SEM studies. Different functions such as movement, stimulus reception, absorption, phagocytosis and secretion are known or suspected to reside in differently specialized surface structures.

Before presenting the micrographs, it seems worthwhile to define the types of cell projections as they are highly variable in structure and function. Some cells have large processes which decide the whole shape of the cell. Neurites and dendrites of neurons, tapering processes of fibroblasts and reticular cells, pseudopods of wandering cells and attenuated skirts of endothelial cells are examples. Besides these main processes or *processes of the first order* there may occur different sorts of *microprocesses, processes of a second order,* which usually measure less than 1 μm in thickness and may cover the processes of the first order. Filopodia, lamellipodia, microvilli, cilia, etc. are included in the category of microprocesses whose structure and function will be reviewed briefly.

Microprocesses of Cells

1. Filopodia

Filopodia were first described as "fibrous projections" by Porter, Claude and Fullam (1945). They are slender, smooth projections either straight or curled in course, occasionally surpassing 5 μm in length. They may be branched and may end in a knob-like (Rajaraman *et al.,* 1974) or, occasionally a palm-like structure as shown in Plate I-2B.

In cultured cells filopodia occur mainly at the cell margin and mostly extend to touch adjacent cells and the substrate (glass) surface (Plate I-2A). Thus, filopodia seem to serve as cell anchors. This view is supported by the fact that they are markedly increased in number when the cell is rounded up for mitosis.

In many cells filopodia are believed to serve as tentacles for detecting objects around the cell. When cells migrate or extend their processes, filopodia may be issued in front and detect the way by a rapid "scanning" movement (Albrecht-Buehler, 1976). Moreover, filopodia play an essential role in the recognition and capture of foreign bodies in macrophages as demonstrated *in vitro* (Parakkal *et al.,* 1974) and *in situ* (Muto and Fujita, 1977; Plate I-2B).

2. Lamellipodia

These correspond to the thin flaps of cells long known by light microscopists as "ruffles" which show a waving movement and are involved in the pinocytotic activity of the cell. Lamellipodia are variable in size but fairly uniform in thickness (100 nm). They may stand on the cell like sails but usually extend from the cell margin. Lamellipodia have been observed mainly in cultured cells (Porter and Fonte, 1973; Plate I-2A), but typical lamellipodia are known to occur in some cells *in situ* like Kupffer cells (macrophages) in the liver (Muto *et al.*, 1977).

Lamellar microprojections of smaller but more uniform size may densely cover the cell surface as in Plate I-2A and Figure I-3. Gradations between these and microvilli may be recognized. This type of microprojection probably differs in function from the typical lamellipodia aforementioned and may better be called *microlamellae*.

3. Globular or bulbous microprocesses (zeiotic blebs)

Globular microprojections measuring 200—500 nm in diameter may occur on the cell surface, often densely grouped. While observing cultured cells, Price (1967) called these microprojections zeiotic blebs (zeiosis: cell bubbling). The function of zeiotic blebs is not clear. Macrophages are often covered by this type of microprojection (Fig. I-1), and there are gradations between typical zeiotic blebs and globules with a haft (knobstick) (Plate I-1). The blebs on macrophages are believed to serve, besides unknown functions, as reservoirs of cell membranes, as they are smoothed out when the cell has phagocytotically internalized voluminous materials and a large area of membrane has been necessary to wrap them (Muto and Fujita, 1977).

4. Cilia

Motile cilia are distributed so widely that we may find them covering certain protozoa and, simultaneously, the epithelial cells of the human airway and oviduct. Under the SEM they are pliantly curved columns with a smooth surface and mostly tapered only in the short segment close to their rounded end; in occasional cases, however, they end rather bluntly.

SEM studies using protozoa (*Paramecium*) have demonstrated that ciliary coordination and the form of ciliary beat can be correctly preserved by suitable fixation and drying techniques (Horridge and Tamm, 1969; Tamm, 1972), and this possibility is being applied to the analysis of ciliary movement in human and animal tissues (e.g., brain ventricles: Yamadori, 1978).

5. Single cilium

Single cilium may be possessed by almost every category of cells, including endocrine cells, neurons, glial cells and mesenchymal cells (cf. Beertsen *et al.*, 1975), although their occurrence at the center of the luminal surface of certain epithelial cells is best known (Plates II-3B, V-8).

Reflecting the construction of their microtubules which usually are more or less reduced in number or deviated in arrangement as compared to the (9+1) pattern in the motile cilia, single cilia under the SEM tend to appear thinner, tapered or uneven in thickness and may be irregularly twisted. In certain cells the single cilium, or its modified forms, serves as the receptor site of the cell, but in many others its function is unknown and may seem to be an "evolutionary remnant" (Latta *et al.*, 1961). Single cilium may be *paired* as demonstrated in Figure I-2.

6. Microvilli

Microvilli are columnar projections of the cell with variable dimension, arrangement and function. The most typical forms may be microvilli covering the intestinal epithelium. Here, individual microvilli are beautifully uniform in dimension, measuring 1—1.5 μm in length and 80 nm in thickness.

Microvilli generally are thought of as a device to increase the cell surface area, thus elevating the efficiency of material transport through the cell membrane. Recent studies have revealed that the intestinal microvilli *actively move* by the actin-myosin mechanism (Mooseker and Tilney, 1975) and evidence is accumulating to indicate that many other types of microvilli are also motile. Moreover, certain types of microvilli, such as those of gustatory cells, gut endocrine cells and mast cells are believed to serve as the *receptor site* of the cell. An interesting experiment using SEM by Wasserman *et al.* (1977) demonstrates that cytochalasin B induces unusual cleavage of the mouse oocyte *in vitro*, in which one cell is covered by microvilli while its partner is naked. The microvillous cell is covered by *lectins*, carbohydrate-binding proteins suspected to mediate cell recognition and adhesion, whereas the naked one is lacking in these substances. The role of microvilli in the agglutination of cells has been demonstrated by Ukena and Karnovsky (1977).

As was the case also in zeiotic blebs, microvilli may serve as the *reservoir of cell membranes.* In their SEM observation of synchronized mastocytoma cells, Knutton *et al.* (1975) demonstrated that in the late G_1 stage there was an accumulation of microvilli which unfolded during mitosis apparently to cover the surface area of new cells.

SEM studies have demonstrated the occurrence of unexpectedly numerous and uniform microvilli on different kinds of cells, including mucus secreting cells (Plates I-3, VII-6, 7), gall bladder epithelium (Plate III-28), mesothelial cells (Andrews and Porter, 1973) and some endothelial cells (Plates II-1A). The "surface anatomy" of lymphocytes shown by SEM has indicated that immunologically important differentiation of B and T cells might be possible by the distribution and shape of their microvilli (Polliack *et al.*, 1973).

It seems worthwhile mentioning in this connection that the "microvilli" of lymphocytes are often quite deviated from the columnar shape and may show *spiny* or *filopodium-like* forms (Fig. I-1). Another instance of a deviated form of microvilli may be seen in Fig. V-4, which demonstrates *lamellar microvilli* found in a proximal tubule of the kidney. *Knobsticks* characteristic of macrophages (Plates I-1,4) are usually treated as modified microvilli.

Stereocilia (Plate I-5) are very long microvilli (over 10 μm) and often branched in their course. They grow densely on the epididymal duct epithelium. Stereocilia, as this early designation indicates, do not beat as cilia but they may possibly be involved in a slow movement. Their mechanical or chemical action upon spermatozoa is not known.

More complicatedly modified microvilli may be called *complex microvilli.* An example of this will be the microvilli covering the chorionic villi of the placenta demonstrated in Plate I-6B.

7. Microplicae

Winding ridge-like folds measuring 100—200 nm in width and 200—800 nm in height occur on the surface of some stratified squamous epithelia lining the oral cavity, pharynx and esophagus (Plate I-7, Fig. I-5). Recently these folds were studied precisely and termed microplicae by Andrews (1976), who proposed that the grooves formed by the ridges might function to hold a layer of lubricating and cushioning mucin designed to protect the epithelium from abrasive abuse.

The microplicae of this type are formed as intercellularly gearing ridges while the cells are flattened and elevated towards the surface of the epithelium.

There are other types of microplicae whose formation mode and functional significance are unknown. Figure I-3 demonstrates an instance found on the pyloric epithelium, while Plate V-8C shows a more typical microplicae characterizing the dark cells of the renal tubules.

8. Microfoliae

Tongue-like microprojections as seen in the epidermal cells of the human sole (Plate I-8) may be best termed microfoliae. This unique surface structure seems to be a device for deep and fast interdigitation of cells against mechanical abuse.

Plate I-1 Microprocesses of Macrophage. Human Spleen.

A macrophage is seen wandering through the space of the splenic cord. In contrast to other cells with a smooth surface, the macrophage is densely covered by microprojections of various forms. Some are blebs and others are long strings and shorter sticks. Some microvilli have a round swelling at the tip.

Certain of these microprocesses, if not all, are believed to serve as chemical and/or physical tentacles used in the locomotion and phagocytosis of this cell. Moreover, the microprojections apparently represent a membrane reservoir as the macrophage may be able to phagocytose enormous amounts of foreign bodies or cells — even 20 erythrocytes at a time —, and for this a large area of cell membrane must be provided partly to cover the surface of the swollen cell and partly to form the phagosomes by membrane invagination. As a matter of fact the macrophage becomes gradually smooth surfaced as voluminous foreign bodies are internalized.

X 10,000

(Reproduced from T. Fujita: *Arch. histol. jap.* **37**: 187—216, 1974).

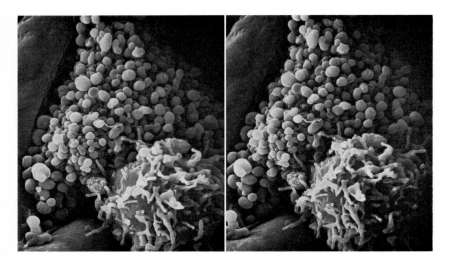

Fig. I-1 Macrophage and Lymphocyte in Stereo. Human Spleen.
A macrophage covered with blebs and knob-sticks is associated with a lymphocyte characterized by horn-like modification of the microvilli. The connection of macrophage and lymphocyte as shown here is believed to represent the process of information transmission from the former to the latter.

X 3,300

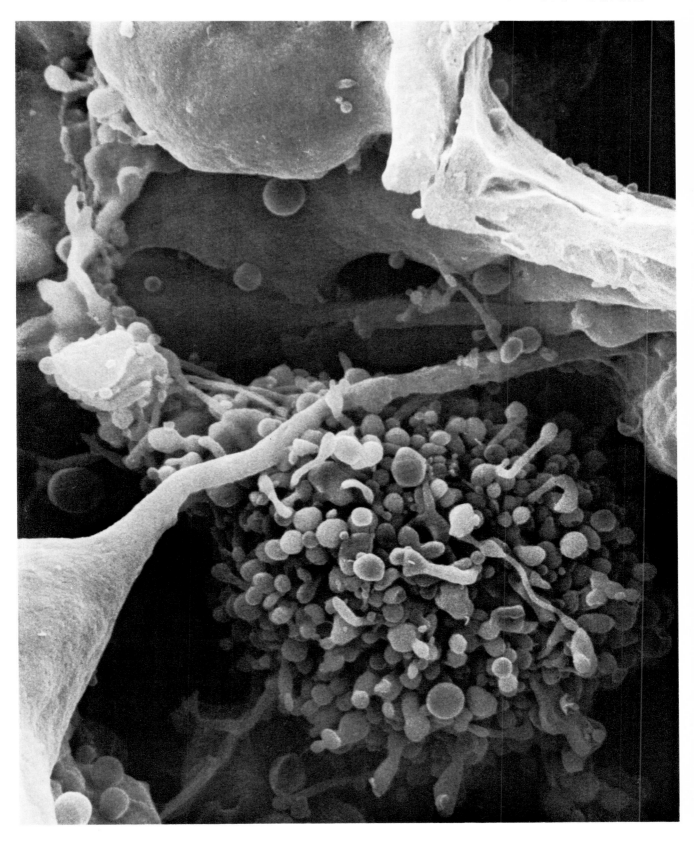

Plate I-2
A. Filopodia and Lamellipodia. Rat Macrophage.

This macrophage cultured *in vitro* clearly shows filamentous microprojections called *filopodia* and web-like ones called *lamellipodia* along the cell margin. The upper surface of the cell is covered by microprocesses *(microlamellae)* which are bulbous or the shape of twisted plates.
   X 8,700

B. Filopodium Catching a Foreign Body. Rat Liver.

One critical role of filopodia may be visualized by this micrograph. A Kupffer cell (macrophage residing in hepatic sinusoid) extends a filopodium which has recognized an erythrocyte previously fixed for a few minutes by glutaraldehyde (Muto and Fujita, 1977). This initial figure of *phagocytosis* was obtained by fixing the liver 2 min after introduction of the fixed erythrocytes into the hepatic circulation.
   X 16,000

(Plate I-2A: Courtesy of Dr. M. Muto, Department of Anatomy, Niigata University School of Medicine)

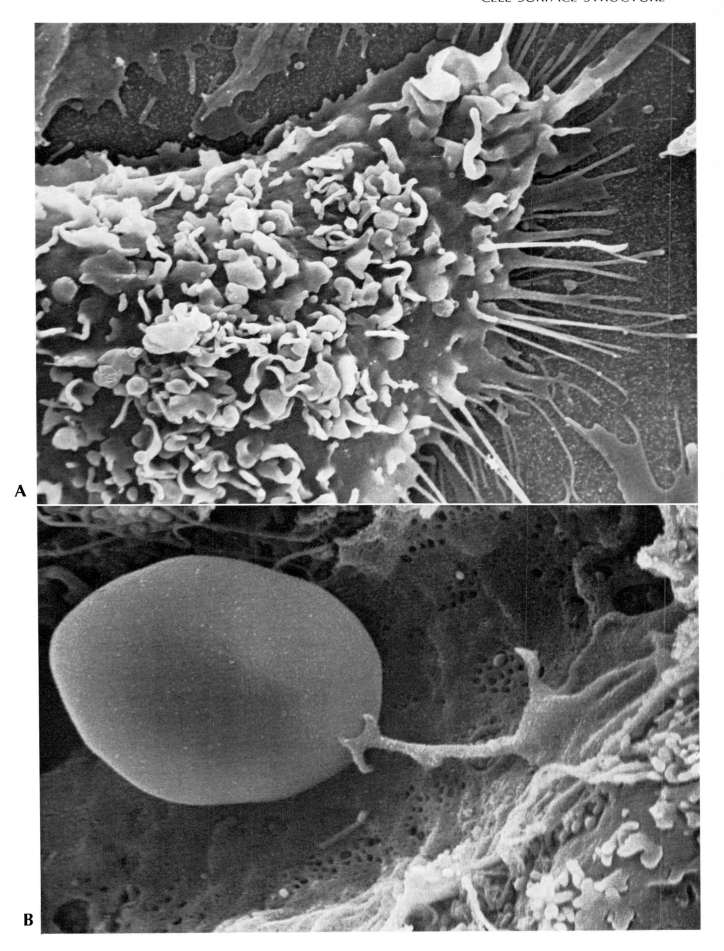

Plate I-3
A. Cilia and Microvilli.   Rabbit Oviduct.

The oviduct epithelium is comprised of two types of cells. One is *ciliated* and the other is *non-ciliated* mucous cells.

The cilia are slender, gently waving columns with a smooth surface, though minute particulate contaminants may often be attached. A few cilia end rather bluntly while majority are tapered near the tip.

Microvilli covering the mucous cell are much smaller than cilia. Presumably they may serve as tentacles to detect chemical or mechanical stimuli to the cell.

   X 7,900

B. Cilia.   Rat Trachea.

The respiratory tract is covered by cilia whose beating movement occurs like waves in the direction from the lung to the nose. Under the SEM the curved shape of the cilia suggests the beating movement and its direction.

   X10,000

C. Cilia and Microvilli.   Human Ductulus efferens.

The ductus efferens of the epididymis is lined by ciliated and non-ciliated cells. The ciliated cell has very long cilia, measuring about 10 μm, and microvilli of irregular length. The non-ciliated cell is covered by longer and more uniform microvilli. Notice that these microvilli are densely dotted with sugar coat substances, whereas the cilia and microvilli of the ciliated cell are smooth surfaced.

   X 12,000

(Courtesy of Dr. K. Goto, Department of Urology, Niigata University School of Medicine).

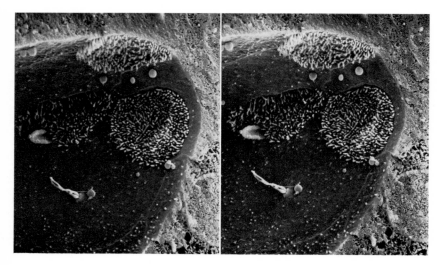

Fig. I-2 Single Cilium in Stereo.   Collecting Tubule of Rat Kidney.
This tubule is lined by cells with cilium and a few cells with microvilli. The cilium demonstrated here incidentally is doubled in structure.

   X 2,700

A

B

C

Plate I-4   Microvilli.   Proximal Convolution of Rat Kidney.

This micrograph shows the luminal view of the microvilli forming the brush border of the proximal tubule. Among the uniform, finger-shaped projections, however, one may find larger ones which are mostly twisted and end in a round swelling. The latter type of microvilli do not belong to the epithelial cells, instead they are issued by a macrophage or macrophages residing in the epithelium and apparently detecting objects to be phagocytosed. (A similar macrophage is shown in fractured view in Figure V-7A).

Tiny, spiny matters covering the microvilli correspond to the sugar coat, fixed and dried. It seems noteworthy that the spines are much more conspicuous on the microvilli of macrophages than on those of the epithelial cells in agreement with the TEM findings that the sugar coat of macrophages is markedly thick.

  X 32,000

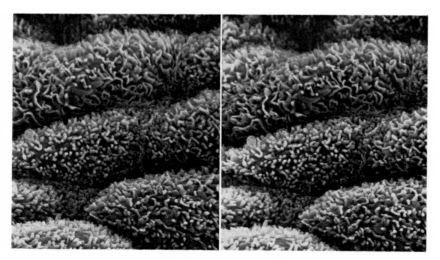

Fig. I-3 Mucosal Surface of Pyloric Antrum in Stereo.   Dog.
The epithelial cell in the upper portion of this micrograph is covered by microlamellae (page 12), while the cell in the middle, by short microvilli. There may be seen gradations between the laminae and villi, especially on the cell in the lower right. Cell boundaries are recognized by the marginal folds of adjacent cells.

  X 4,600

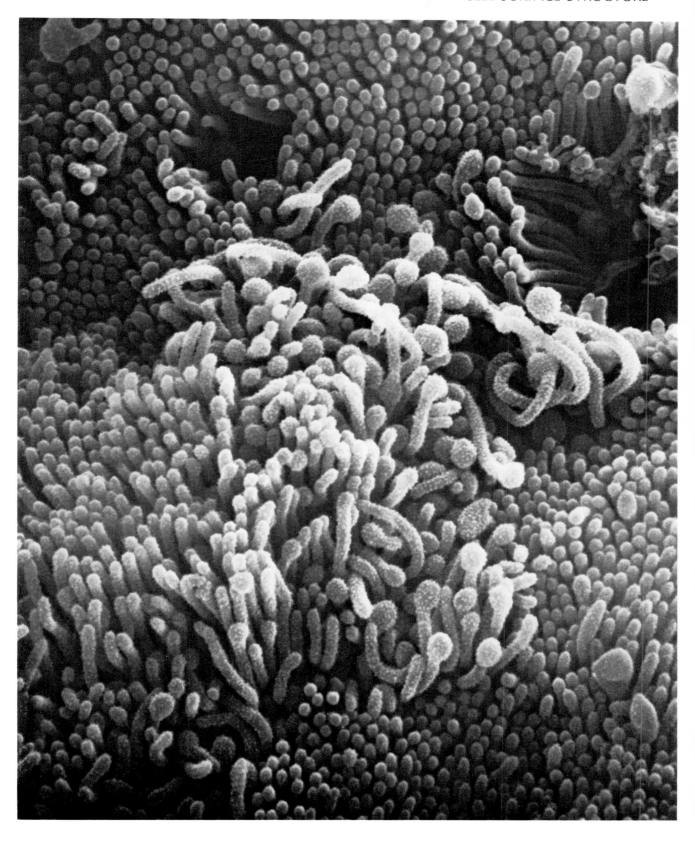

Plate I-5
A. Stereocilia, Luminal View.   Epididymal Duct, 65 Year-Old Man.

Stereocilia thickly covering the epididymal duct are viewed from the lumen. The stereocilia appear rather irregular in length and direction. As seen in this micrograph, they usually touch spermatozoa or even cover them. The globular bodies presumably represent secretions of the epithelium.
   × 7,000

B. Stereocilia, Slide View.   Epididymal Duct, 73 Year-Old Man.

The basal portion of stereocilia reveals that these long microvilli (40 μm in this specimen) form brush-like tufts, each of which originates from a cytoplasmic stem projecting at the apex of the epithelial cell. Among the long microvilli there are found shorter ones, which may likely represent newly growing elements. The globules attached to the microvilli presumably are secreted matters.
   × 7,400

(Plate I-5A: Courtesy of Dr. K. Goto, Department of Urology, Niigata University School of Medicine).
(Plate I-5B: Reproduced from K. Goto: *Biomed. Res.* 2, Suppl.: 361—374, 1981)

Plate I-6
A. Processes of Synovial Membrane Cells.   Rabbit Knee Joint.

The synovial membrane limiting the joint cavity is lined by *mesothelial cells.* These cells project an irregular-shaped cytoplasmic process into the joint cavity. This process, rod-like and frequently branched, issues thread-like microprojections.
   X 3,700

B. Complex Microvilli on Chorionic Villi.   Human, 3 Months Pregnancy.

The *syncytial cells* of chorionic villi of the placenta are densely covered by microvilli, which often are complex in structure as this micrograph demonstrates. From an axial microvillus radiate numerous secondary microvilli, which may further branch. The end of each projection is characteristically swollen.
   X 13,000

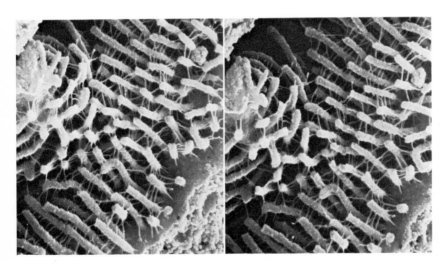

Fig. I-4 Surface Substance in Stereo.   Human Chroionic Villus.
This stereo-pair demonstrates microvilli in a cryptal space incidentally found in the syncytial layer of a 3 month placenta. Apparently representing the coagulated surface material of the cell, delicate filaments radiate from and connect individual microvilli. In his classical TEM study of gall bladder epithelium, Yamada (1955) first described the image of what today is called the sugar coat under the designation of *antennulae microvillares.* The unusually clear filaments shown here may well deserve this designation.
   X 16,000

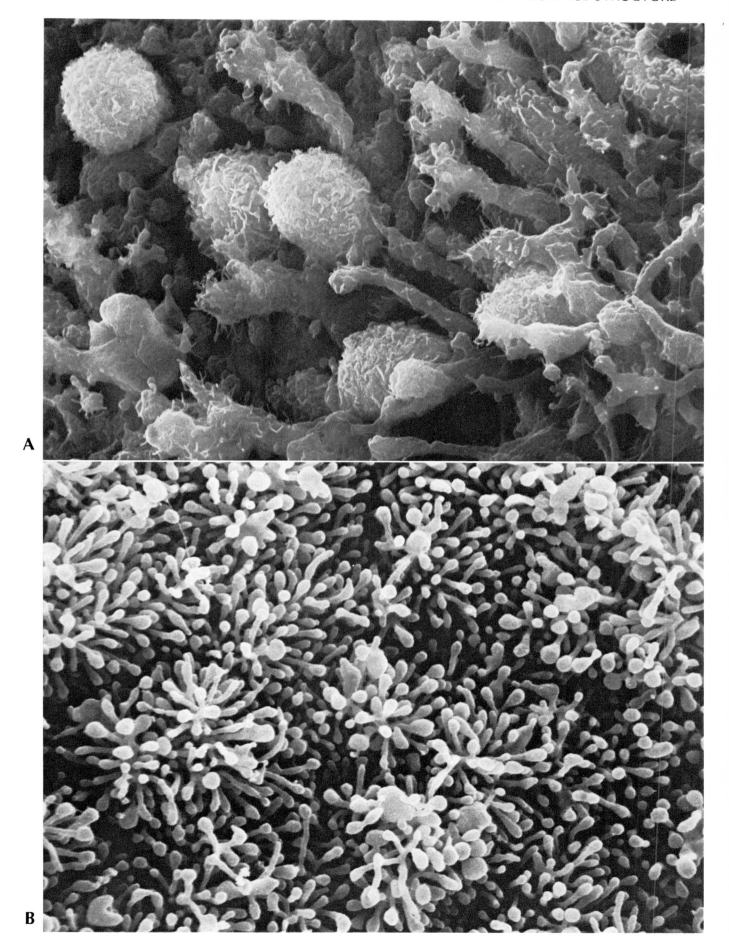

Plate I-7   Microplicae.   Rabbit Esophagus.

The squamous epithelial cells of the esophagus are flattened polyhedral in shape and their surface is covered with twisted ridges or microplicae. In most parts the ridges form numerous circles, while in smaller portions they are extended in parallel lines.

In closer view, the microplicae partly are simple crests of the cell surface but partly appear to be formed by columnar cytoplasm attached to the cell surface. As a matter of fact, these "columns" seem to be lifted up in a few places and apparently continue to microvilli which occur sporadically. It is, at least, worthwhile noticing that the "columns" of the microplicae and microvilli are identical in dimension and surface structure.

   X 12,000

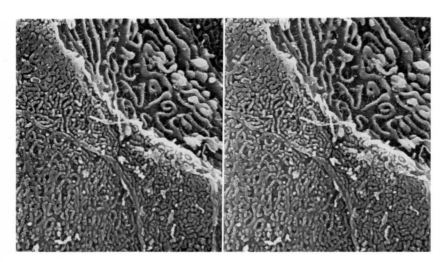

Fig. I-5 Microplicae in Stereo.   Rabbit Esophagus.
A squamous cell shows its different facets with different patterns of microplicae. The upper right portion which is a facet exposed by removal of an overlying cell shows real "plicae" forming parallel, circular and irregular patterns. On the luminal facet, which occupies the major part of this micrograph, the microplicae are smaller in size and denser in distribution, and their gradation to microvilli is likely.

   X 3,700

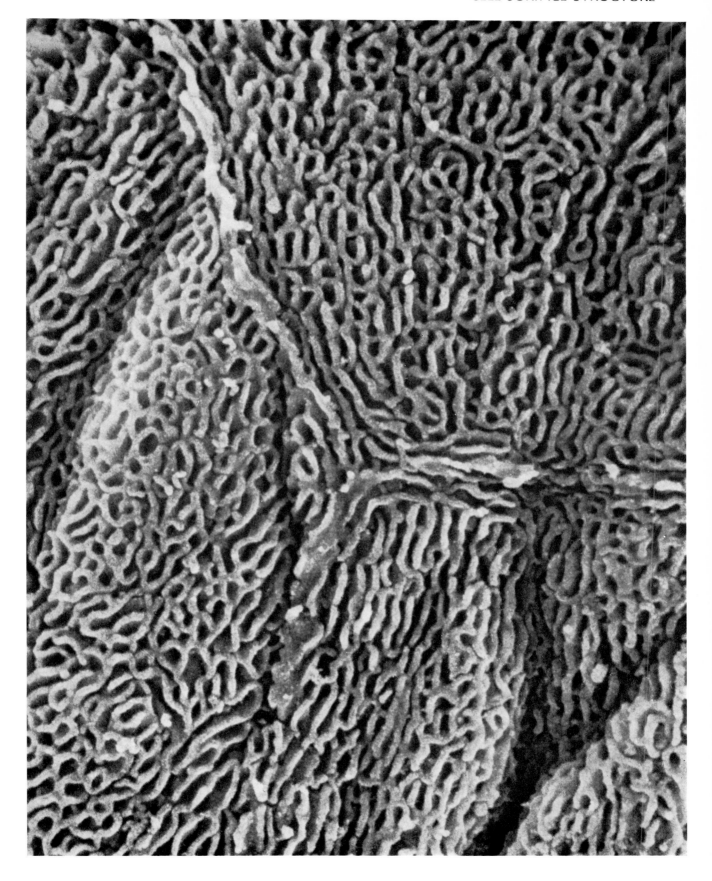

Plate I-8   Microfoliae.   Squamous Epithelium of Human Sole.

The squamous epithelial cells of the *sole* are also flattened polyhedral forms and their facets are covered by scale-like projections. These microfoliae are most conspicuously developed in the skin of this part of the body and are geared to the corresponding grooves of adjacent cells, thus preventing sliding between the thickly piled up cells.

In closer view (**B**), the microfoliae are tongue-like in shape and smooth surfaced. Their end occasionally is divided.

A: X 3,300,   B: X 23,500

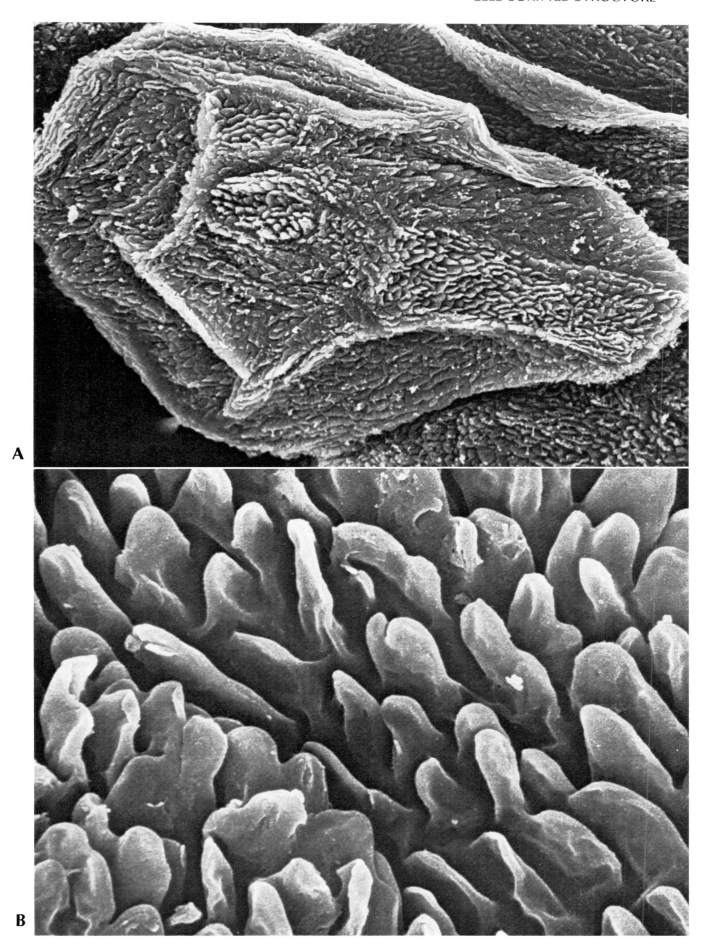

Cell Internal Structure

The internal structure of cells became the subject of SEM study from 1970 along with the initiation of different kinds of *cracking methods* (Haggis, 1970; Makita and Sandborn, 1971; Nemanic, 1972; Tanaka and Iino, 1972; Tanaka, 1974). The main difficulty thereby exists in that the organelles to be observed are embedded in the cytoplasmic matrix and are difficult to dissect selectively. In some conditions or techniques of cracking, fracture may occur along the membranes and boundaries of subcellular structures, thus revealing their plastic contours. *Ion etching* (page 6) may be applied to expose some intracellular structures relatively resistant to ion bombardment (Fig. I-6, Plate I-12A, Plate I-13B) (Fujita *et al.*, 1974; Tanaka *et al.*, 1976a). Treatment of cells with hypotonic solutions has been shown to be effective for some purposes (Tanaka *et al.*, 1977). More recent method by Tanaka (1981) is a checmical etching technique by the use of dilute OsO_4 (page 6; Plate 1-10).

Most of the SEM studies of cell internal structures still have been visualization of organelles and their arrangement already established by TEM. However, there are new subjects properly suited to SEM observation, like the three-dimensional analysis of *chromatin fibers* (Tanaka and Iino, 1973) and the study of *nuclear pores* (Kirschner *et al.*, 1977). Many other structures are believed to be waiting for creative use of SEM combined with the invention of new preparation techniques.

Plate I-9   Fracture of Cell.   Rat Liver Cell.

This freeze-cracked liver cell gives an overview of one of the typical arrangements of cell organelles. At the top the nucleus (N) reveals its convex surface as the fracture occurred along the nuclear membrane.

The cytoplasm is fractured more flatly in this case. It is divided into compartments for different organelles. The major compartments are filled with lamellar stacks of *rough endoplasmic reticulum* (ER). Smaller parts are occupied by oval and elongate profiles of mitochondria. The loose collections of coarse granules (G) probably correspond to glycogen areas.

The materials were treated with a hypotonic solution before fixation according to Tanaka *et al.* (1977).

X 32,000

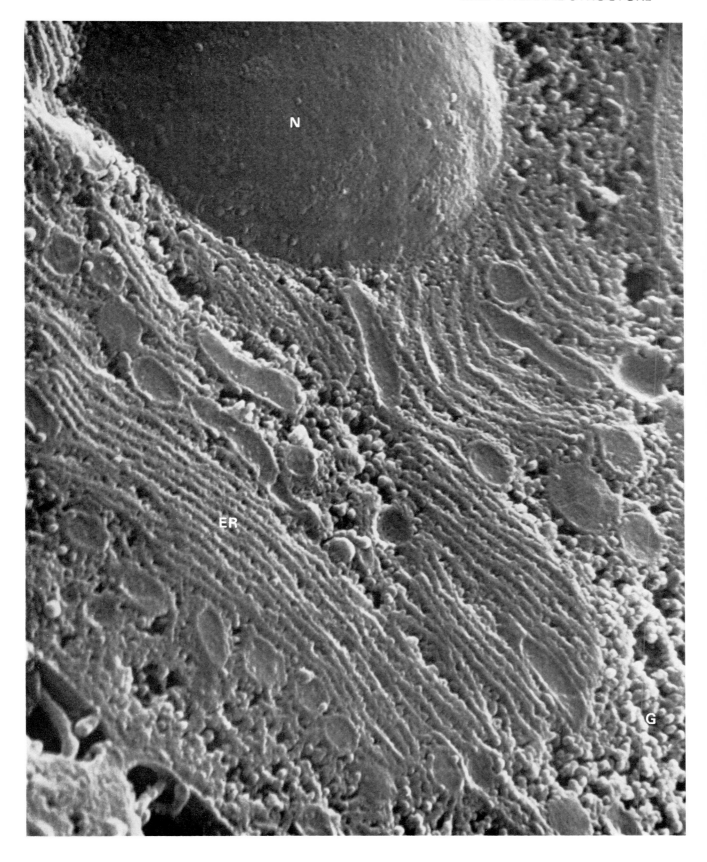

Plate I-10 Golgi Complex. Rat Epididymis.

By the use of the Tanaka's maceration technique with dilute OsO$_4$ (page 6), a Golgi complex was exposed from the cytoplasmic matrix. The Golgi complex consists of a stack of flattened lamellae studded with vesicles of a fairly uniform size and a network of tubules developing on the inner side (trans face) of the lamellae. The tubule network contains granules of various sizes which probably correspond to lysosomes.

   X 69,000

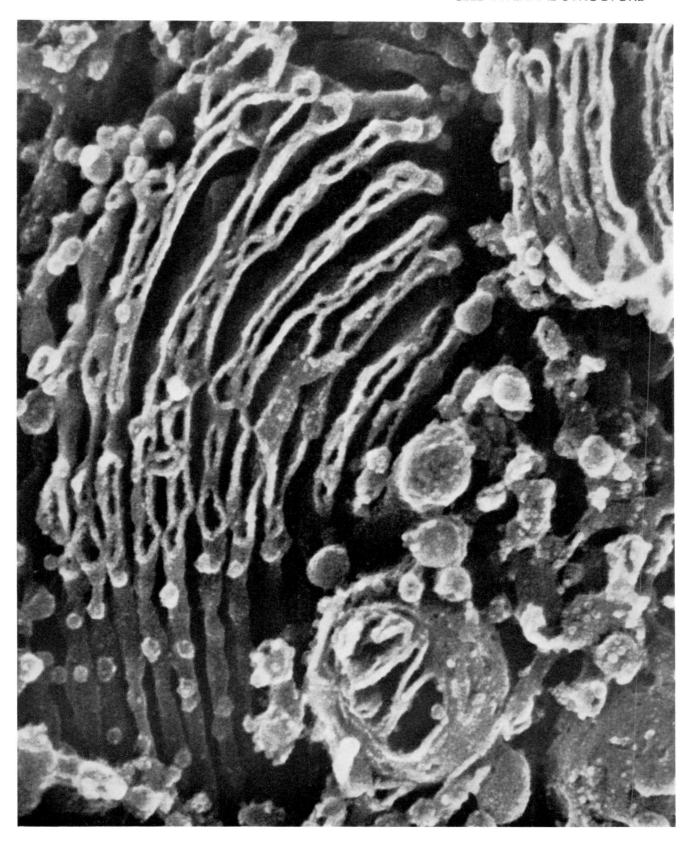

Plate I-11

A. Nuclear Envelope.   Acinar Cell of Dog Pancreas.

The nucleus or karyoplasm shows clumpy and fibrous aggregations of granular matter. The fibrous forms may possibly be based on the occurrence of chromatin fibers in their core (Tanaka and Iino, 1973).

The nuclear envelope consists of an *inner* (I) and an *outer leaflet* (O). The inner leaflet appears rather smooth and shows small perforations corresponding to *nuclear pores*. The outer leaflet is densely covered with granules probably corresponding to ribosomes at least in the major part. *Nuclear pores* are seen surrounded by several granules, forming rosette-like *"pore complexes"* (Kirschner *et al.*, 1977; Schatten and Thoman, 1978).

In the lower right one may see saccular cisterns of *endoplasmic reticulum* with the *rough surface* structure seen on the outer nuclear membrane.

X 48,000

B. Nuclear Pores.   Rat Liver Cell.

This high power micrograph of the *outer leaflet* of the nuclear envelope shows several *nuclear pores* surrounded by the *pore complex* mentioned above. The rosette granules of the complex stand out in this specimen as other, ribosomal granules have been mostly lost by previous treatment of the tissue with a hypotonic solution (Tanaka *et al.*, 1977).

X 190,000

C. For legend see page 36

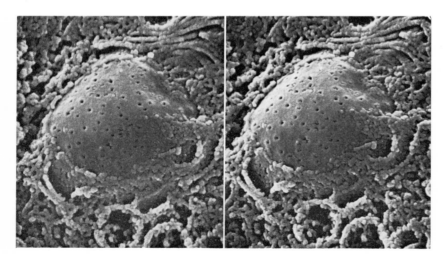

Fig. I-6 Nuclear Enverope in Stereo.   Ion-Etched Dog Pancreas Cell.
When the cracked surface is slightly ion-etched, *nuclear pores* on the inner leaflet are accentuated. The outer leaflet is covered with granules and nuclear pores are not evident here.

X 14,000

(Plate I-11A and Fig. I-6: Reproduced from K. Tanaka *et al.*, *Arch. histol. jab.* 39: 165–175, 1976)

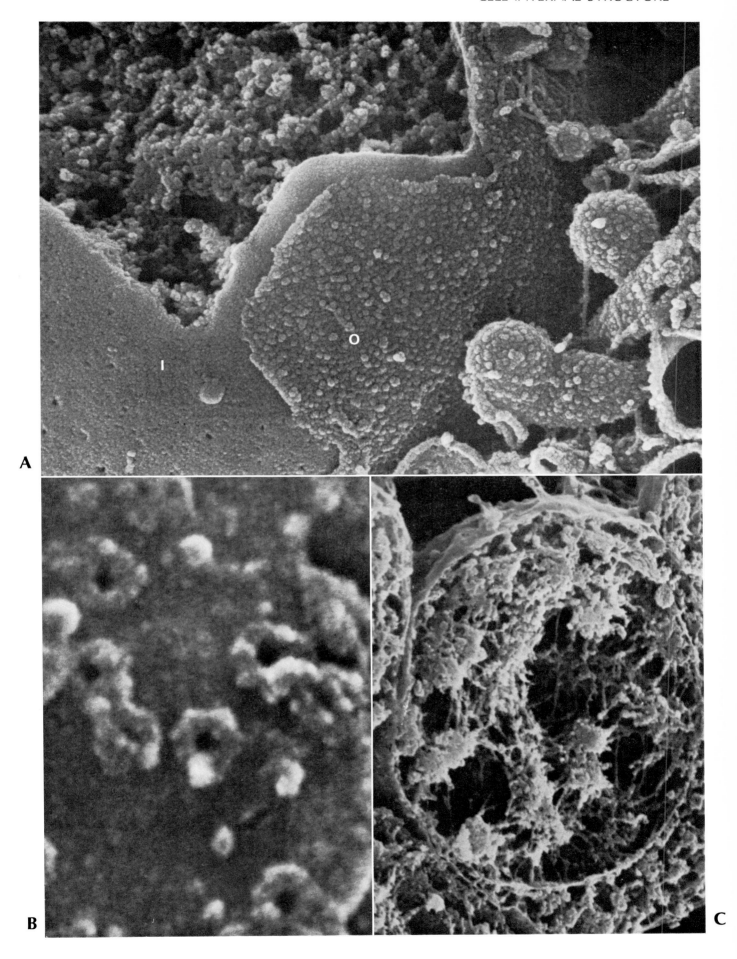

Chromosomes are the carriers of genes, but little has been elucidated concerning their fine structure. From the results of TEM studies it is likely that the chromosome is constructed of supercoiled or folded chromatin fibers about 10 nm in thickness. SEM is advantageous over TEM for observation of such coiled structures.

Plate I-11   C. (Preceding Plate)
Nuclear Division, Early Prophase. Grasshopper *(Pachytilus danicus)* testis.

Nuclear Division, Early Prophase. Grasshopper (Pachytilus danicus) Testis.
In this nucleus chromosomes are already formed which consist of *chromatin fibers* about 25 nm thick. Individual chromosomes are connected by delicate *interchromatide fibers*.
   Fixation in 0.5% OsO_4; ethanol cracking without etching.
   X 9,200

Plate I-12
A. Giant Chromosomes.   Salivary Gland of Drosophila Larva.

These are the largest known chromosomes among animals. The *cross striations* of the chromosomes, long known by light microscopy, are evident in this SEM image. Further detailed structures of the chromosomes are regrettably obscure in this picture.
   Styren resin cracking; ion-etching.
   X 12,000

B. Chromatin Fibers of Giant Chromosomes.

After treatment of drosophila cells with ATA (aurin tricarboxylic acid) according to the method of Tsutsui *et al.* (1975), the nucleus swells up and the fibrous structures of chromatin fibers become clear. Chromatin fibers forming the giant chromosomes are thus visualized. Individual fibers are 25 nm in thickness and show cross striation with a pitch of ca 10 nm (Tanaka *et al.*, 1976b).
   X 115,000

C. Human Chromosomes.   From Cultured Blood Lymphocyte.

Human blood lymphocytes cultured for three days were treated with a special hypotonic solution (Iino, 1974, 1975), fixed in Carnoy's fluid and air-dried.
   This X-shaped chromosome consists of a network of fibers, 70—100 nm thick. Surrounding the chromosome are seen more delicate filaments 30 nm in thickness apparently coming loose from the chromosomal fibers. These filaments show crossbands of 25 nm periodicity.
   X 16,500

(Plate I-12: Reproduced from A. Iino: *Acta anat. nippon.* 49: 337—344, 1974)

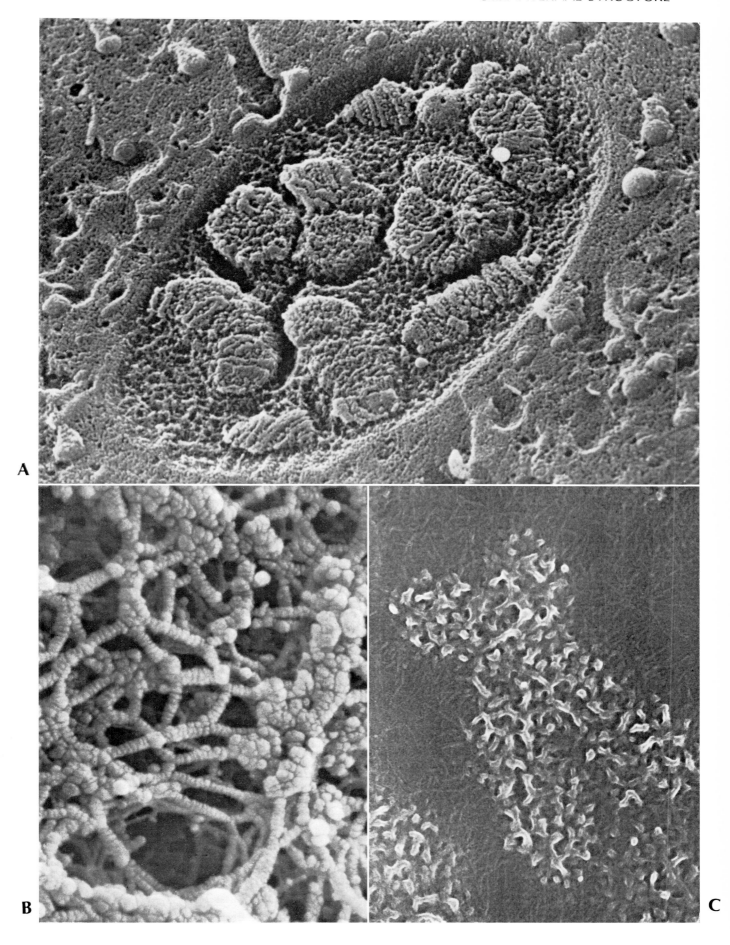

Plate I-13
A. Lamellar Endoplasmic Reticulum.   Anterior Horn Cell of Cat Spinal Cord.

The lamellar cisterns of endoplasmic reticulum are demonstrated with their smooth inner surface. The spaces between the cisterns are filled with granular matter which may reasonably be regarded as *ribosomes*. The whole image thus represents a typical Nissl body structure.

    The specimen, after styren resin cracking, was slightly ion-etched.

    N: nucleus.

    × 84,000

B. Tubular Endoplasmic Reticulum.   Acinar Cell of Rabbit Pancreas.

Tubular endoplasmic reticulum, after ion-etching, reveals its branching extensions.

    N: nucleus.

    × 56,000

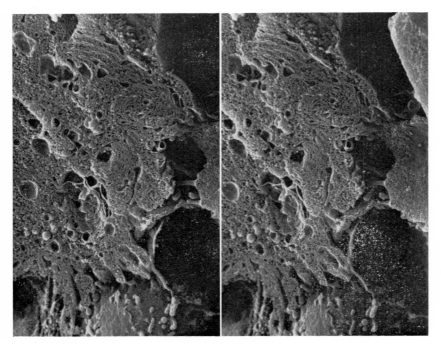

Fig. I-7 Golgi Complex of Macrophage in Stereo.   Dog Lymph Node.

A Golgi complex consisting of lamellae, vacuoles and vesicles is three-dimensionally observed. Some vesicles contain round granules which probably represent *lysosomes*. A lamellar stack of rough endoplasmic reticulum is seen at the top, and one of its cisterns seems to be connected with a vacuole which also belongs to a Golgi complex.

    × 6,400

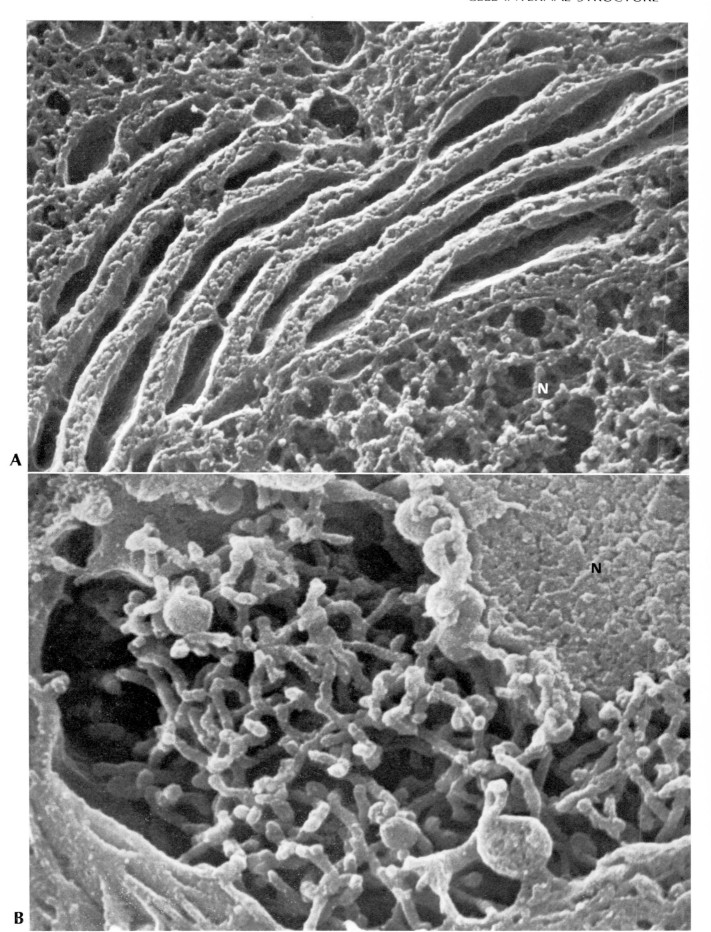

Plate I-14

A. Mitochondria.   Rat Cardiac Musle Cell.

Mitochondria are one of the cell organelles whose identification by SEM is often difficult. There may occur different kinds of granules and bodies in the cell which resemble mitochondria in shape.

In this specimen, which was freeze-cracked in isoamyl acetate and received no ion-etching, round mitochondria are identified as such by their clearly exposed *cristae*.

X 27,000

B. Connected Mitochondria.   Rat Cardiac Muscle Cell.

Based on the TEM studies of serial sections it has been demonstrated in fungi (Hoffmann and Avers, 1973), protozoa (Paulin, 1975) and even in mammals (Rancourt *et al.*, 1975), that the cell may have a single huge mitochondrion of reticular extension instead of numerous mitochondria.

In this micrograph, some mitochondria are shown connected by constricted portions. It is, however, unknown whether such a continuous structure of mitochondria prevails or is occasional, or whether it is constant or transient in occurrence.

X 56,000

C. Isolated Mitochondria.   Japanese Macaque Adrenocortex.

Mitochondria isolated from adrenocortical cells by Schneider's method using MSTE medium were embedded in epoxy resin and cracked. *Cristae mitochondriales* reveal their fractured profiles and, partly, their *en face* view.

Tubular instead of *lamellar* cristae are generally seen as the mitochondrial inner structure of adrenocortical and other steroid-producing cells. In monkeys, however, many adrenocortical cells, especially those of the zona glomerulosa, have been demonstrated by TEM studies to possess lamellar cristae in their mitochondria (Brenner, 1966; Penny and Brown, 1971).

X 48,000

(Plate I-14: Courtesy of Prof. T. Shimada, Department of Anatomy, Ohita University Medical School).

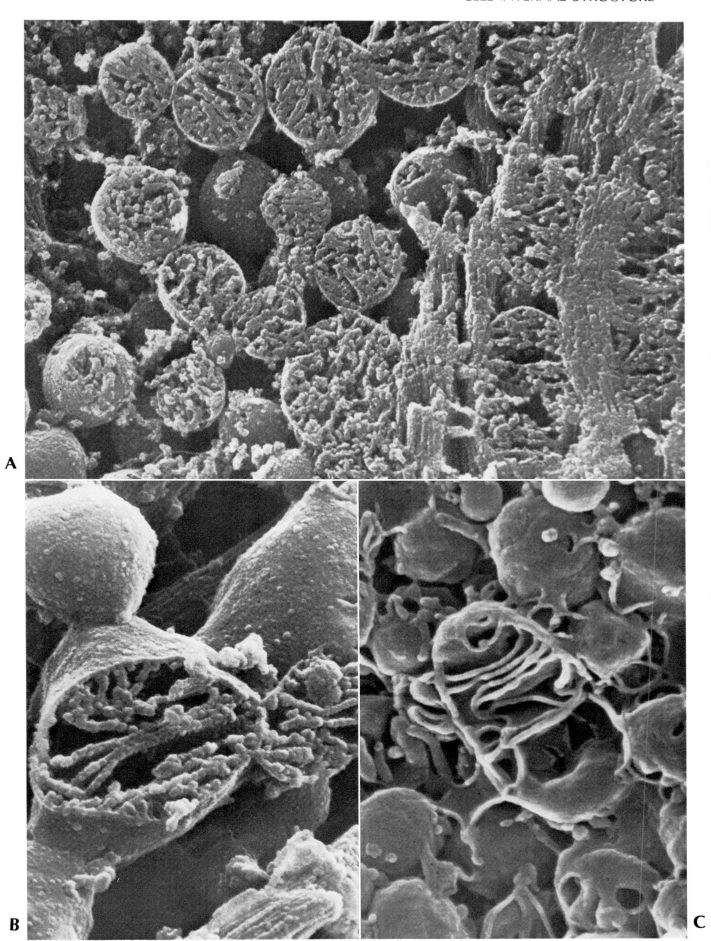

Plate I-15

A. Secretary Granules and Microtubules. Cat Pancreatic Acinar Cell.

In a fractured pancreatic acinar cell globular secretory granules are exposed even without ion-etching. Filaments of about 45 nm in diameter are found which are attached to the surface of secretory granules.

From their dimension and distribution in the cell, the filaments may be reasonably judged as *microtubules*. The banding on their surface might be ascribed to the subunits of microtubules which have been proposed by Tilney *et al*. (1973) on the basis of TEM findings.

The relation of microtubules to secretory granules has attracted the attention of cell biologists as there is a hypothesis that some microtubules are attached to the granules and, by contraction, cause movement of the latter towards the cell surface where they are released (Lacy, 1970; Poisner, 1970).

X 156,000

B. Crystalloid Cores of Insulin Granules. Dog Pancreatic B Cell.

As an example of paraplasmas observed by SEM, this micrograph shows the granule cores of an islet B cell which are known to contain insulin. In the dog the crystalloid structure of the granule cores is most conspicuous among mammalian species. They appear as "rods" under the TEM and the real shape has been difficult to analyse by the use of TEM techniques. This SEM view simply and clearly shows that the insulin containing cores are *polygonal plates* with rounded corners (Tanaka *et al.*, 1976a).

The cytoplasm was loosened in this specimen by treatment of the tissue with a hypotonic solution. Freeze-cracking in DMSO; no ion-etching.

X 48,000

(Plate I-15A: Reproduced from Tanaka, K., K. Iino and T. Naguro: *Arch. histol. jap.* **39**: 165–175, 1976)

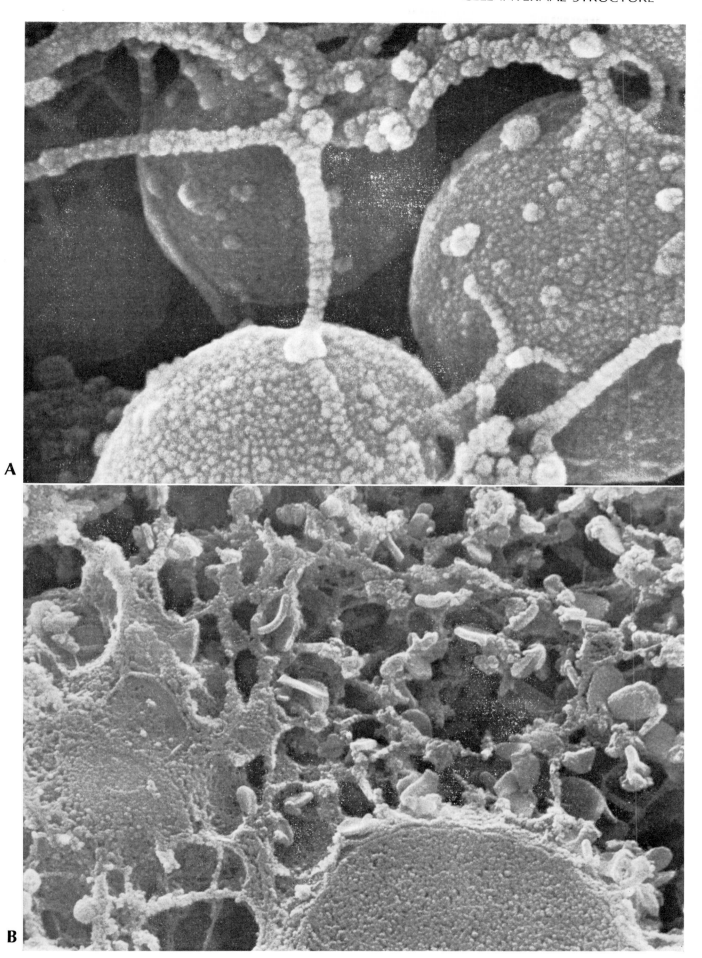

References

Albrecht-Buehler, G.: Filopodia of spreading 3T3 cells. Do they have a substrate-exploring function? *J. Cell Biol.* **69**: 275—286 (1976).

Andrews, P.M.: Microplicae: Characteristic ridge-like folds of the plasmalemma. *J. Cell Biol.* **68**: 420—429 (1976).

Andrews, P. M. and K. R. Porter: The ultrastructural morphology and possible functional significance of mesothelial microvilli. *Anat. Rec.* **177**: 409—426 (1973).

Beertsen, W., V. Everts and J. M. Houtkooper: Frequency of occurrence and position of cilia in fibroblasts of the periodontal ligament of the mouse incisor. *Cell Tiss. Res.* **163**: 415—431 (1975).

Brenner, R. M.: Fine structure of adrenocortical cells in adult male rhesus monkeys. *Amer. J. Anat.* **119**: 429—454 (1966).

Fujita, T., T. Nagatani and A. Hattori: A simple method of ion-etching for biological materials. An application to blood cells and spermatozoa. *Arch. histol. jap.* **36**: 195—204 (1974).

Haggis, G. H.: Cryofracture of biological material. In: (ed. by) O. Johari: Scanning Electron Microscopy/1970. IIT Research Institute, Chicago, 1970 (p.97—104).

Hoffmann, H.P. and C. J. Avers: Mitochondrion of yeast: ultrastructural evidence for one giant, branched organelle per cell. *Science* **181**: 749—751 (1973).

Horridge, G. A. and S. L. Tamm: Critical point drying for scanning electron microscopic study of ciliary motion. *Science* **163**: 817—818 (1969).

Iino, A.: The fine structure of human somatic chromosomes in relation to the pH value of hypotonic medium. *Acta anat. nippon.* **49**: 337—344 (1974).

Iino, A.: Human somatic chromosomes observed by scanning electron microscope. *Cytobios* **14**: 39—48 (1975).

Kirschner, R. H., M. Rusli and T. E. Martin: Characterization of the nuclear envelope, pore complexes, and dense lamina of mouse liver nuclei by high resolution scanning electron microscopy. *J. Cell Biol.* **72**: 118—132 (1977).

Knutton, S., M. C. B. Sumner and C. A. Pasternak: Role of microvilli in surface changes of synchronized P815Y mastocytoma cells. *J. Cell Biol.* **66**: 568—576 (1975).

Lacy, P. E.: Beta cell secretion from the standpoint of a pathobiologist. *Diabetes* **19**: 895—905 (1970).

Latta, H., A. B. Maunsbach and S. C. Madden: Cilia in different segments of the rat nephron. *J. biophys. biochem. Cytol.* **11**: 248—252 (1961).

Makita, T. and E. B. Sandborn: Identification of intracellular components by scanning electron microscopy. *Exp. Cell Res.* **67**: 211—214 (1971).

Mooseker, M. S. and L. G. Tilney: Organization of an actin filament-membrane complex. Filament polarity and membrane attachment in the microvilli of intestinal epithelial cells. *J. Cell Biol.* **67**: 725—743 (1975).

Muto, M. and T. Fujita: Phagocytotic activities of the Kupffer cell: A scanning electron microscope study. In: (ed. by) E. Wisse and D. L. Knook: Kupffer Cells and Other Liver Sinusoidal Cells. Elsevier, Amsterdam, 1977 (p. 109—119).

Muto, M., M. Nishi and T. Fujita: Scanning electron microscopy of human liver sinusoids. *Arch. histol. jap.* **40**: 137—151 (1977).

Nemanic, M. K.: Critical point drying, cryofracture, and serial sectioning. In: (ed. by) O. Johari and I. Corvin: Scanning Electron Microscopy/1972. IIT Research Institute, Chicago, 1972 (p.297—304).

Parakkal, P., J. Pinto and J. M. Hanifin: Surface morphology of human mononuclear phagocytes during maturation and phagocytosis. *J. Ultrastr. Res.* **48**: 216—226 (1974).

Paulin, J. J.: The chondriome of selected trypanosomatids. A three-dimensional study based on serial thick sections and high voltage electron microscopy. *J. Cell Biol.* **66**: 404—413 (1975).

Penney, D. P. and G. M. Brown: The fine structural morphology of adrenal cortices of normal and stressed squirrel monkeys. *J. Morphol.* **134**: 447—466 (1971).

Poisner, A. M.: Release of transmitters from storage: A contractile model. *Adv. biochem. Psychopharmacol.* **2**: 95—108(1970).

Polliack, A., N. Lampen, B. D. Clarkson and E. De Harven: Identification of human B and T lymphocytes by scanning electron microscopy. *J. exp. Med.* **138**: 607—624 (1973).

Porter, K. R., A. Claude and E. F. Fullam: A study of tissue culture cells by electron microscopy. *J. exp. Med.* **8**: 233—246 (1945).

Porter, K. R. and V. G. Fonte: Observations on the topography of normal and cancer cells. In: (ed. by) O. Johari and I. Corvin: Scanning Electron Microscopy/1973. IIT Research Institute, Chicago, 1973 (p.683—688).

Price, Z. H.: The micromorphology of zeiotic blebs in cultured human epithelial (HEp) cells. *Exp. Cell Res.* **48**: 82—92 (1967).

Rajaraman, R., D. E. Rounds, S. P. S. Yen and A. Rembaum: A scanning electron microscope study of cell adhesion and spreading in vitro. *Exp. Cell Res.* **88**: 327—339 (1974).

Rancourt, M. W., A. P. McKee and W. Pollack: Mitochondrial profile of a mammalian lymphocyte. *J. Ultrastr. Res.* **51**: 418—424 (1975).

Schatten, G. and M. Thoman: Nuclear surface complex as observed with the high resolution scanning electron microscope. Visualization of the membrane surfaces of the nuclear envelope and the nuclear cortex from *Xenopus laevis* oocytes. *J. Cell Biol.* **77**: 517—535 (1978).

Tamm, S. L.: Ciliary motion in paramecium. A scanning electron microscope study. *J. Cell Biol.* **55**: 250—255 (1972).

Tanaka, K.: Frozen resin cracking method and its role in cytology. In: (ed. by) M. A. Hayat: Principles and Techniques of Scanning Electron Microscopy. 1, Nostrand Reinhold, New York, 1974 (p. 125—134).

Tanaka, K. and A. Iino: Frozen resin cracking method for scanning electron microscopy and its application to cytology. Proc. 30th Ann. Meeting Electron Microsc. Soc. Amer.: 408—409 (1972).

Tanaka, K. and A. Iino: Demonstration of fibrous components in hepatic interphase nuclei by high resolution scanning electron microscopy. *Exp. Cell Res.* **81**: 40—46 (1973).

Tanaka, K., A. Iino and T. Naguro: Scanning electron microscopic observation on intracellular structure of ion-etched materials. *Arch. histol. jap.* **39**: 165—175 (1976a).

Tanaka, K., A. Iino, T. Naguro and H. Fukudome: Scanning electron microscopic study on chromatin fibres and chromosomes. *J. Electron Microsc.* **25**: 212 (1976b).

Tanaka, K., K. Ozawa, T. Naguro and Y. Kashima: Hypotonic solution treatment on biological materials for observing intracellular structures by scanning electron microscope. *J. Electron Microsc.* **26**: 251 (1977).

Tilney, L. G., J. Bryan, D. J. Bush, K. Fujiwara, M. S. Mooseker, D. B. Murphy and D. H. Snyder: Microtubules: Evidence for 13 protofilaments. *J. Cell Biol.* **59**: 267–275 (1973).

Tsutsui, K., M. Yamaguchi and T. Oda: Aurintricarboxilic acid as a probe for the analysis of chromatin structure. *Biochem. biophys. Res. Commun.* **64**: 493–500 (1975).

Ukena, T. E. and M. J. Karnovsky: The role of microvilli in the agglutination of cells by concanavalin A. *Exp. Cell Res.* **106**: 309–325 (1977).

Wasserman, P. M., T. E. Ukena, W. J. Josefowicz, G. E. Letourneau and M. J. Karnovsky: Cytochalasin B-induced pseudocleavage of mouse oocytes in vitro. II. Studies of the mechanism and morphological consequences of pseudocleavage. *J. Cell Sci.* **26**: 323–337 (1977).

Yamada, E.: The fine structure of the gall bladder epithelium of the mouse. *J. biophys. biochem. Cytol.* **1**: 445–458 (1955).

Yamadori, T.: Scanning electron microscopic studies of the ciliary beat on the wall of the brain ventricles and spherical structures on the wall of the central canal. In: (ed. by) R. P. Becker and O. Johari: Scanning Electron Microscopy/1978 II. Scanning Electron Microscopy, Inc., AMF O'Hare, 1978 (p. 823–830).

CHAPTER

CIRCULATORY SYSTEM AND HEMATOPOIETIC TISSUES

Blood Vessels

Endothelial Microprocesses

SEM observation of vascular endothelia has revealed that the luminal surface of some larger as well as smaller blood vessels, including capillaries, may be covered with microvillous, hair-like and granular *microprocesses* (Smith *et al.*, 1971; Tokunaga *et al.*, 1973; Edanaga, 1974; Albert and Nayak, 1976; Fujita *et al.*, 1976) (Plates II-1, 3A). Although it has been claimed that the endothelial microprojections may be increased in number and length by rinsing with warm saline for several minutes prior to fixation (Edanaga, 1974; Peine and Low, 1975), the occurrence of microvilli apparently is not so much artifactual as intrinsic to the site and nature of the vessels. In the rabbit, for instance, *numerous microvilli* were seen in the sinus aortae (Plate II-3A), sinus trunci pulmonalis and aorta ascendens, whereas they were absent in the arcus aortae (Plate II-3B), aorta thoracica and their main branches as well as in the truncus pulmonalis; the smooth surfaced endothelia of the latter group of vessels were usually provided with single cilia (Edanaga, 1974). *Single cilia* are common on the endocardial cells at least in the rabbit (Edanaga, 1975).

The functions of the endothelial microprojections are unknown. It is worth stressing that in many, if not all, blood vessels the endothelial surface is *not* smooth as previously believed and the vascular rheology should be largely revised according to this fact. Furthermore, the endothelial *surface area* in contact with blood is enormously larger than previously thought.

Classification of Blood Capillaries and Small Vessels

Mainly based on the SEM observations of the architecture of the endothelial lining, blood capillaries and other small vessels are classified as follows:

1) *Closed type* in which the endothelial pavement shows no perforations (Plates II-1, 2, 3). This type, which also covers larger blood vessels, is also called the *muscular type* as the capillaries in muscular tissue were thought to represent this type. It has been pointed out, however, that capillaries in skeletal muscle (rat) may be partly fenestrated (Korneliussen, 1975).

2) *Pored* or *fenestrated type* in *endocrine, renal* and some other tissues (Plate II-4A). Thin and pored compartments of the endothelial cytoplasm, called areolae fenestratae (Fujita *et al.*, 1976), are bordered by cytoplasmic crests arborizing from the nuclear swelling.

3) *Large fenster type* or *hepatic type* (Plate II-4B). A few gaps occur intercellularly but the majority are represented by intracellular fensters, some of which are small and form areolae fenestratae as in the endocrine type but others are large, sometimes exceeding 1μm in diameter.

46

Although Wisse (1970) proposed in his TEM study in the rat that only the endocrine type pores are natural structures, later SEM studies in different species have confirmed the occurrence of larger fensters which could not be regarded as mere artifacts (Motta and Porter, 1974; Motta, 1975, Muto, 1975; Grisham et al., 1976; Muto et al., 1977).

4) *Intercellularly gapped* or *lattice type* seen in the *splenic sinus* (Plates II-4C, 9). The specialized endothelium is formed by parallel rod cells connected by their side processes, between which spindle-shaped spaces are open and allow migration of cells (Miyoshi et al., 1970; Fujita, 1974; Suzuki et al., 1977).

5) *Reticular type* occurring in *postcapillary venules.* Thick endothelial cells of stellate shape possessing side processes are interwoven in a reticular sheet and only determined lymphocytes can migrate through the spaces between the cell processes (Umetani, 1977) or though the cell bodies (Cho and De Bruyn, 1979).

Comments and micrographs pertaining to the second to fifth type blood vessels will also be found in the chapter on each organ concerned.

Adventitial Aspect of Blood Vessels

The adventitial aspect of the cells forming the vascular wall had been hidden under the SEM because of the dense basal lamina and fibrous elements covering them, until the method by Evan (1976) opened the way to *denude the cells* by the use of warmed hydrochloric acid and collagenase after fixation of the tissue. Stimulated by this method, Uehara and Suyama (1978) devised a method to digest the tissue with collagenase and trypsin prior to fixation.

By the use of these methods the outer aspect of capillary endothelal cells, the dendritic shape of pericytes (Murakami et al., 1979) (Plate II-5) and the three-dimensional arrangement of muscle cells in the tunica media (Uehara and Suyama, 1978) have been clearly demonstrated under the SEM. Visualization under the SEM of autonomic nerves supplying the vascular wall seems an interesting field to be explored.

Plate II-1   Blood Capillary in Gut Mucosa.   Dog.

A branching capillary of closed type is shown in the lamina propria of canine duodenum. The vessel partly reveals its endothelial outside covered by a felt of reticular fibers but its major part is longitudinally open showing its luminal surface. Marginal folds are inconspicuous but *microvilli* are condensed along the cell margins so that the boundaries of endothelial cells may be indicated. Besides microvilli, round bodies of presumable cytoplasmic nature are seen on the endothelial surface. The endothelial cells lining a vessel are not always uniform in structure, presumably and partly because of wear and tear and the renewal of the cells. A cell with a rough and lamellated surface structure (*) is shown in this picture; the significance of this structure is unknown.

The *lamina propria* represents a typical reticular tissue. Cells and thin processes, probably including nerve fibers, are loosely embedded in a meshwork of reticular fibers.

X 4,000

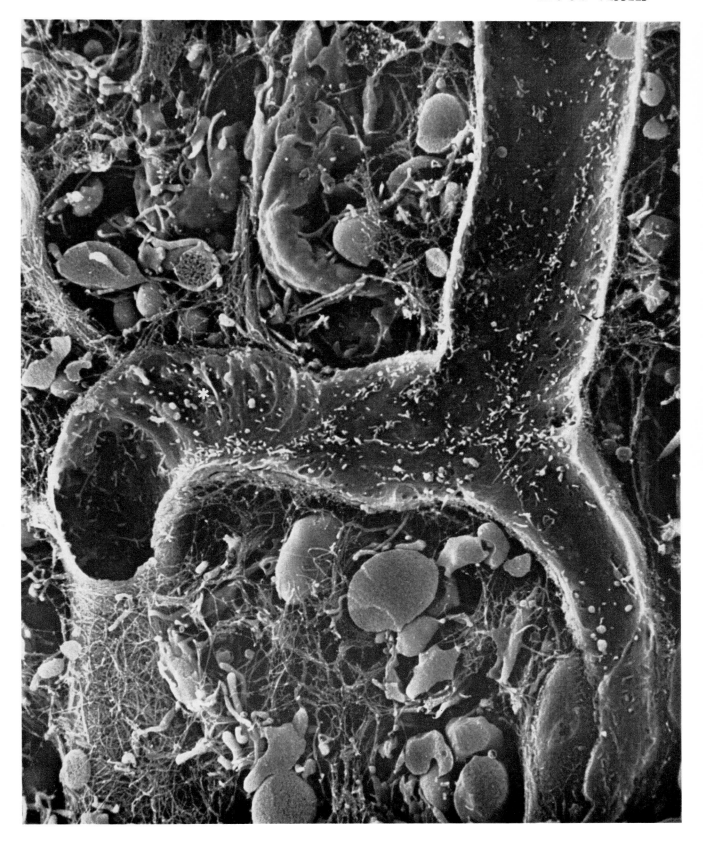

Plate II-2   Endothelium of Small Vein.   Rabbit Uterus.

Endothelial cells lining veins generally are polygonal in outline and a round swelling is formed by the nucleus. In this micrograph the vascular wall is edged because of branching in the upper left, and around the edge the endothelial cells are elongated in shape.

The cell border is fringed with a *marginal fold* known to occur in many types of blood vessels (Fawcett, 1963). Short microvillous and granular projections are dispersed on the cell surface. Larger granular and vesicular matter seems to be partly contamination but partly cell microprocesses.

X 2,600

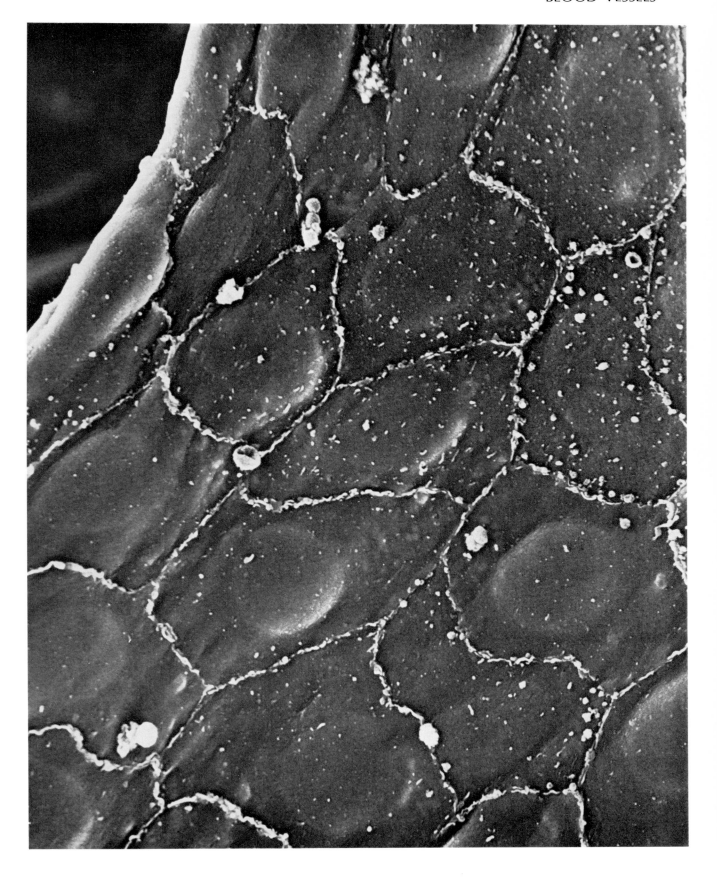

Plate II-3
A. "Haired" Endothelial Cells. Aortic Sinus of the Rabbit.

Endothelial cells whose nuclear portion is domed are densely covered with microvillous or hair-like projections of cytoplasm. They are twisted, irregular in length and may measure 2–3 μm. According to the results of an experiment by Edanaga (1974), the endothelial hairs of rabbit aorta are recognized after perfusion of the vessel with warm (37–40°C) saline, but may become shortened into inconspicuous granular projections after perfusion with cold (2°C) saline.
    Arrow indicates an endothelial cell boundary showing marginal folds.
    X 7,000

B. Ciliated Endothelial Cells. Aortic Arch of the Rabbit.

Endothelial cells are elongated in shape and smooth surfaced in this micrograph. Each of the nuclear protrusions possesses a *single cilium*. As far as the adult rabbit is concerned, single cilia of endothelial cells have been found in the aortic arch, thoracic aorta and some portions of the heart (Edagana, 1974,1975).
    Arrows indicate endothelial cell boundaries.
    X 9,000

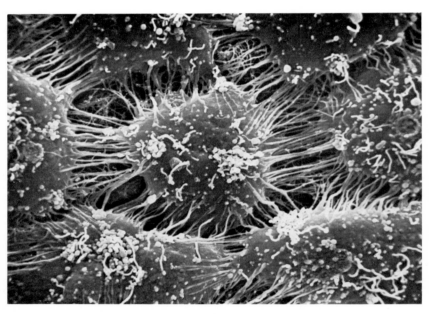

Fig. II-1 Juvenile Endothelial Cells. Aortic Sinus of Newborn Rabbit.
In juvenile animals the endothelial pavement may still be incomplete. The cells are connected with each other by numerous microprojections (filopodia) radiating from the margin of the cell, leaving spaces through which the underlying fibrous structures of intimal tissue can be seen. This type of cell connection seems to reflect the *mesenchymal nature* of endothelial cells in this stage of development.
    The upper surface of endothelial cells is covered by hair-like, drumstick-shaped and granular microprojections. The occurrence of single cillium is not clear in this micrograph.
    X 3,800

(Plate II-3A, 3B and Fig. II-1: Courtesy of Dr. M. Edanaga, Research Laboratories of Yoshitomi Pharmaceutical Industries, Ltd.)

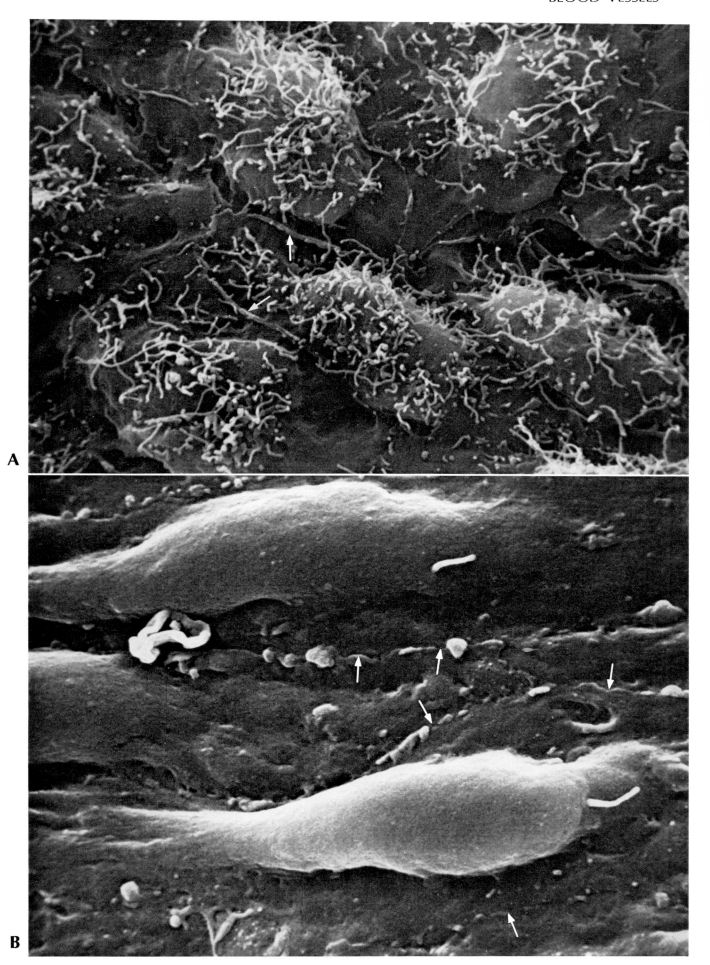

Plate II-4   Four Specialized Forms of Endothelium.

As reviewed on page 46, we classify the endothelial structure of blood vessels into five types. The first, closed type endothelium has been demonstrated in Plates I-1, 2 and 3 and the other four which are more specialized in structure and function are gathered on the opposite page.

A is a *pored type* capillary from rat glomerulus. In this micrograph the areolae fenestratae bordered by cytoplasmic crests are evident. Arrows indicate endothelial boundaries.

B demonstrates the *large fenster type* of endothelium from rat hepatic sinusoid. The arrow indicates the cell boundary.

C shows the *intercellularly gapped type* or *lattice type* of endothelium from the human splenic sinus. A detailed account of this micrograph can be found on pages 64—71.

D is the *reticular type* from a postcapillary venule of rabbit tonsil.

A: × 11,000,   B: × 4,400,   C: × 2,200,   D: × 6,500

(Plate II-4B: Courtesy of Dr. M. Muto, Department of Anatomy, Niigata University School of Medicine).
(Plate II-4D: Courtesy of Dr. Y. Umetani, Department of Otolaryngology, Niigata University School of Medicine).

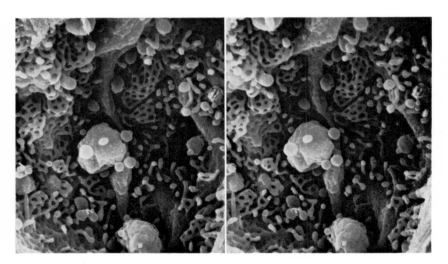

Fig. II-2 Active Formation of Endothelial Microprojections in Stereo.
Rat Renal Glomerulus.

In normal adult rats occasional portions of glomerular capillaries may show unusually numerous microprojections. Conspicuously enough, plates and baskets of the pored structure are projecting into the lumen. Pored plates seem to develop from microvillous forms. This activated microprojection formation may presumably represent a process of the renewal of pored endothelial sheet.

× 8,500

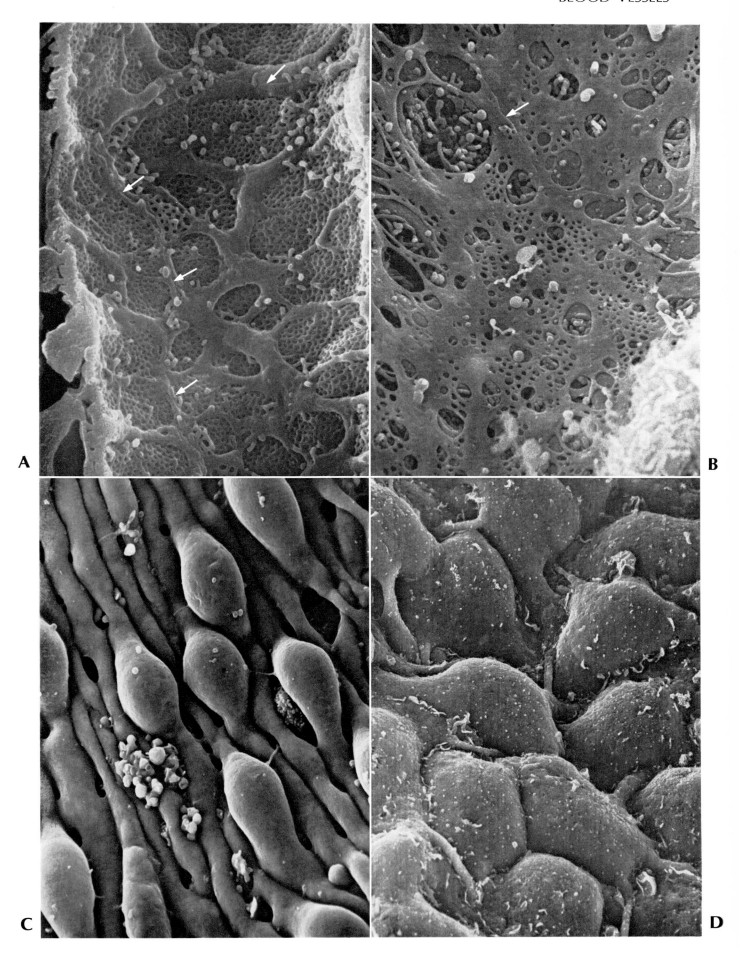

Plate II-5   Pericytes of Blood Capillaries.   Capillaries in Rabbit Retina.

The rabbit retina was treated by the HCl-collagenase method of Evan (1976; see page 6) and the outer aspect of the blood capillaries was exposed.

A demonstrates a thin, genuine blood capillary. Two *pericytes* with spindle shaped cell bodies extend long, branching processes, the peripheral portions of which seem to be in tight contact with the endothelium.

B represents an instance of thicker capillaries either close to arterioles or venules. These capillaries are characterized by pericytes of irregular shapes with flattened and branching processes. Notice the complicated distribution of the pericyte processes.

The morphology of pericytes now being revealed by SEM confirms the precise observations by Zimmermann (1923) who left beautiful sketches of the cells stained by Golgi's silver impregnation technique. The function of pericytes is still controversial. Their possible contraction has not been demonstrated in mammalian blood vessels.

A: X 2,300,   B: X 5,900

(Plate II-5A: Reproduced from M. Murakami: *Arch. histol. jap.* 42: 297—303, 1979)
(Plate II-5B: Courtesy of Dr. M. Muarkami, Department of Anatomy Faculty of Medicine, Kurume University).

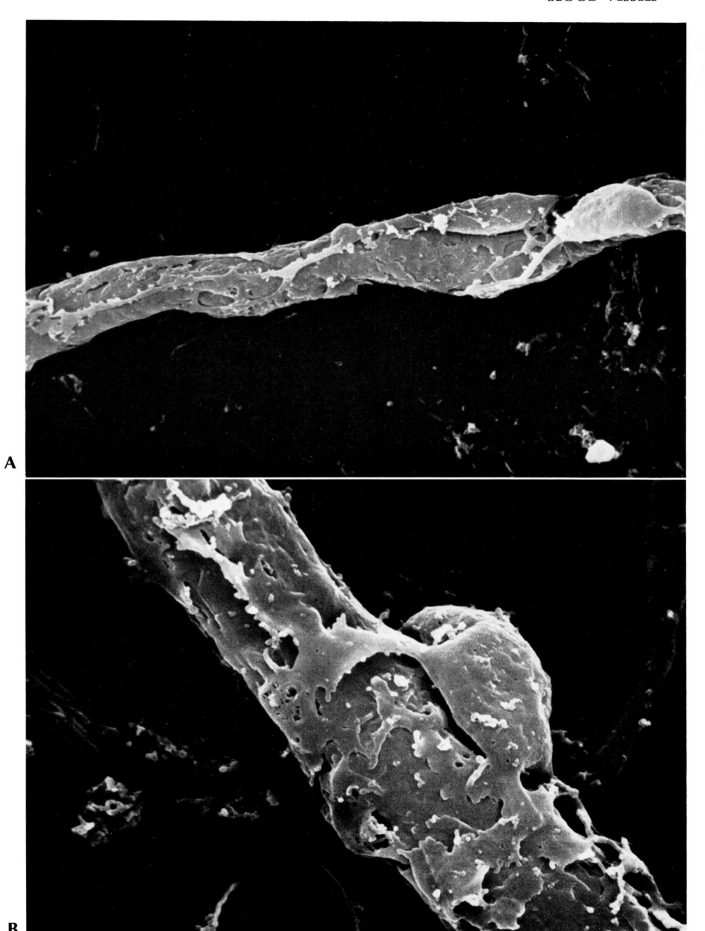

Bone Marrow

The sinus wall is lined by an endothelial sheet incompletely backed by an attenuated cytoplas of so-called *adventitial cells* which apparently are a variant of reticular cells.

The attention of electron microscopists has been focused on the structure of the sinus endothelium as the site of *selective passage* of blood cells matured to a terminal stage in the parenchyme of bone marrow.

Although some earlier TEM studies reported that the sinus endothelium possessed preformed perforations for the passage of blood cells (Zamboni and Pease, 1961; Hudson and Yoffey, 1966; Weiss, 1965, 1970), most of the more recent TEM and SEM investigators agree that the pores seem to occur only in conjunction with migrating blood cells (Campbell, 1972; Leblond, 1973; Becker and De Bruyn, 1976; Muto, 1976).

The mechanism of selection as to how only blood cells of a certain maturity can be allowed to enter the circulation is unknown. While leukocytes actively migrate through the sinus wall, erythrocytes are generally said to move passively through the wall by growth pressure of the extravascular space. On the basis of the strongly constricted figure of erythrocytes passing through a very small pore of the sinus wall, Campbell (1972) postulated that erythrocytes in their reticulocyte stage might possibly be capable of active movement.

Plate II-6   Overview of Bone Marrow.   Rat Femur.

Bone marrow of a young animal is fractured longitudinally. The *central vein* is opened on the left hand and numerous *sinuses* open into it. The vein and sinuses are both lined by very thin endothelium which lacks perforations visible at this magnification. Instead, many *blood cells* are seen attached to the endothelium. One may also see long threads, partly beaded, on the endothelial surface (arrows). These are forming *blood platelets* and will be treated in more detail later (page 91).

The spaces outside of the vessels are filled with bone marrow parenchyme which is the site of *blood cell formation*. Besides small round cells one may see a few large ones which probably correspond to megakaryocytes and adipose cells.

A: a nutrient artery.

X 390

(Micrographs of Plates II-6, 7 and 8 were produced by Dr. M. Muto, Department of Anatomy, Niigata University School of Medicine)

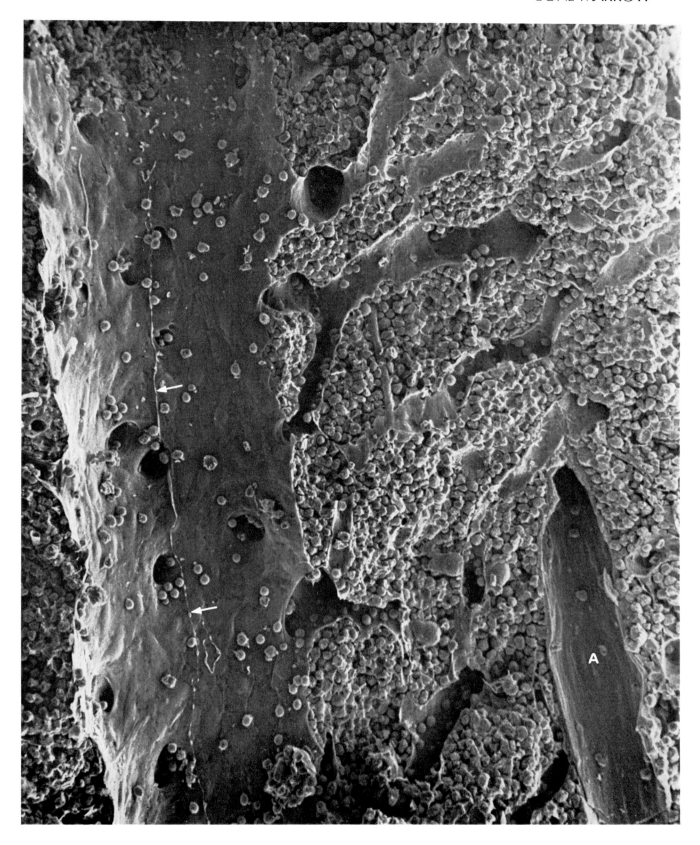

Plate II-7   Bone Marrow Sinus and Blood Cell Migration.   Rat Femur.

A sinus is longitudinally opened and its inside aspect is shown. Except for a few small pores, the apertures of the sinus endothelium are occupied by white (W) and red (R) blood cells (reticulocytes) *passing through* the wall. There are some lymphocytes (L) just attached to the endothelial surface but tightly enough to resist the vascular perfusion, though in this picture it is unknown whether the lymphocytes are merely attached to the wall or hanging on an endothelial hole with a cell constriction.

The outside of the endothelium is partly covered by attenuated precesses of *adventitial cells* (A) which are nothing but reticular cells located close to the sinus. In the meshwork of reticular cell processes different kinds of blood cells undergoing maturation are seen.

Arrows : endothelial cell boundaries.

X 4,000

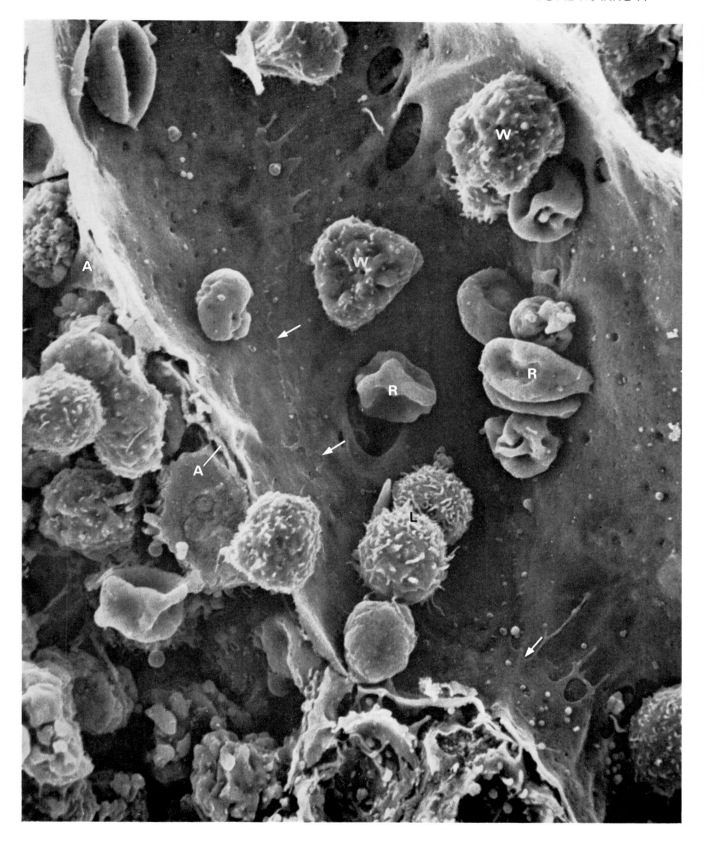

Plate II-8
A. Endothelial Fenestrations of Bone Marrow Sinus.   Rat Femur.

The sinus endothelial cells are very flat on the surface showing few microprojections and marginal folds. Dentate edges of the cells rather tightly overlap and only occasionally a small pore may be left intercellularly.

There are two types of intracellular fenestrations : one is represented by larger apertures (1—3 μm in diameter) and the other by a sieve-like collection of small pores (50—100 μm, arrows). The former are either occupied or touched (as in this micrograph) by migrating blood cells. A few vacant apertures may presumably have been caused by detachment of hanging cells. The significance of the latter type fenestrations is unknown but, as suggested by Becker and De Bruyn (1976), the small pores may likely indicate the initial stage of fenestration for cell migration caused by contact of a blood cell, and larger apertures may be formed by fusion of adjacent pores.

Very small pits covering the endothelial surface apparently are pinocytotic invaginations which are known to be numerous in the sinus endothelial cells (De Bruyn *et al.*, 1975).

X 8,000

B. Bone Marrow Parenchyme.   Rat Femur.

The parenchyme of red marrow is the *blood cell-forming* or *hematopoietic* tissue. Immature erythrocytes (R), mostly corresponding to normoblasts and reticulocytes, are identified by their smooth but characteristically indented surface. One of them (arrow) is under migration through the sinus wall which consists of an endothelial (E) and an adventitial (A) cell. Immature leukocytes occur also numerously; their types are difficult to identify under the SEM with our present knowledge of their surface features.

Rt : reticular cell processes which divide the marrow tissue into compartments.

X 31,000

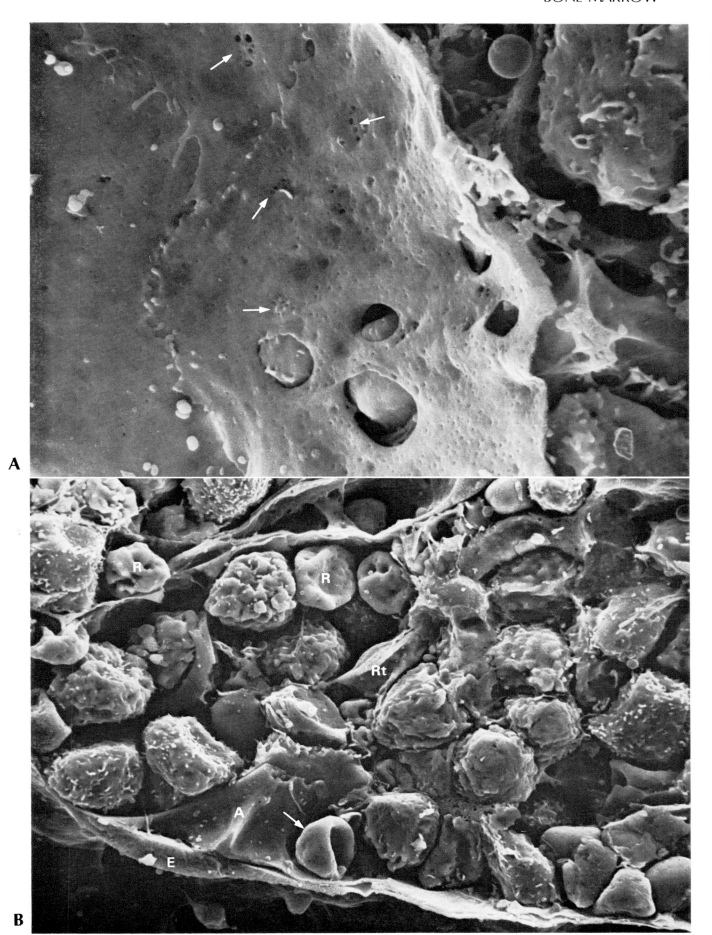

Spleen

The complicated and controversial structure of the spleen has finally become unequivocal and understandable to everybody by observation of perfused organs under the SEM. The spleen is composed of *red* and *white pulp*, of which the former will be dealt with here, as SEM studies have concentrated on it.

The *red pulp* is formed by *splenic sinuses* and *cords of Billroth*. The sinuses are specialized vessels lined by rod cells arranged in parallel and connected by their side-processes. Perfused spleen reveals elongate perforations between these side-connections of rod cells (Miyoshi *et al.*, 1970; Miyoshi and Fujita, 1971; Suzuki, 1972; Leblond, 1973; Fujita, 1974). It is natural that the perforations are widened by the pressure of perfusion, and to a certain extent also during the specimen drying. But of importance is the fact that there are preformed and patent spaces on the sinus wall for the passage of free cells, which only a few TEM investigators have been able to demonstrate (Pictet *et al.*, 1969).

The spongy tissue between the sinuses are the *cords of Billroth. Reticular cells* with wing-like processes form the framework of the cords and also support the wall of the sinuses. The reticular cells are a type of fibroblast, while the *sinus rod cells* presumably are also of the same category of cell but slightly modified towards endothelial cells. Recent TEM studies have revealed the smooth muscle-like filamentous structure in the reticular cells (Saito, 1977) and in the sinus rod cells (De Bruyn and Cho, 1974), thus suggesting the contractile nature of the cells.

Round cells, mainly neutrophils and lymphocytes are seen in the meshwork of reticular cells; blood platelets are numerous.

Macrophages are identified by their characteristically rough surface with numerous microprojections. They are attached to reticular cells in the cords and to rod cells in the sinus. Some cells are rounded in shape but others appear under ameboid migration among the meshwork of cells. Only macrophages are capable of phagocytosing *large* matters like red blood cells, whereas rod cells and reticular cells can take up only small particles. There are no gradations between these cells and macrophages (Miyoshi *et al.*, 1970; Miyoshi and Fujita, 1971; Fujita, 1974).

The problem of whether the arteries of the red pulp, penicillar arteries, are open to the cords of Billroth (*open theory*), or whether they are directly connected with the sinuses (*closed theory*) has been disputed for several decades. Careful observation of perfused spleens by SEM (human: Fujita, 1974; Irino *et al.*, 1977; dog: Suzuki *et al.*, 1977; chicken: Fukuta *et al.*, 1976) has revealed that the arterial terminals are perforated of split to allow blood cell passage into the spaces of the Billroth cords. SEM observations or resin casts of splenic vessels also support the open theory (Irino *et al.*, 1977).

Plate II-9   Red Pulp Overview.   Human Spleen.

The major part of this fracture is occupied by a labyrinth of *sinuses*. The sinuses are lined by a lattice-work of *rod cells* whose oval nuclei swell into the sinus lumen. Perforations in this specialized endothelium are obvious under this magnification. Paired round bodies on the sinus wall are red blood cells hanging in a dumb-bell shape (arrows).

The *cords of Billroth* consist of stellate-shaped reticular cells and different kinds of free cells rounded in shape. Arteries are also contained in the cords. In this micrograph a pair of *sheathed arteries* (Sh) are crossly cut.

Some *macrophages* with rough cell surface are identified already at this magnification either projecting into the sinus lumen or contained in the cordal spaces (∗).

X 1,100

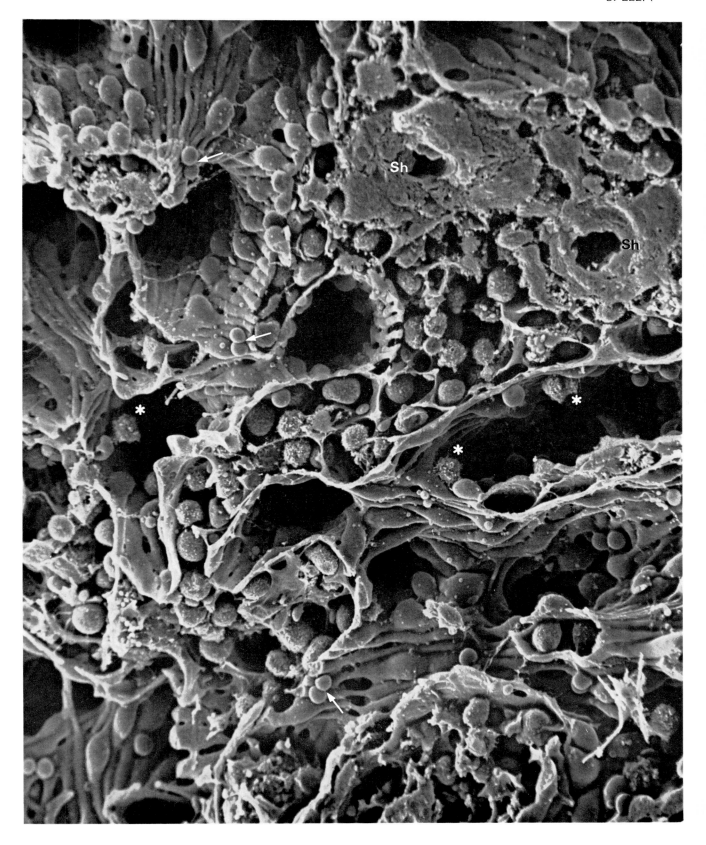

Plate II-10 Cross Section of Sinus and Cord. Human Spleen.

A *sinus* is fractured crosswise on the left hand, revealing the lattice structure of its wall. Rod cells are connected with each other by their *side processes.* Through the apertures between the rod cells, processes of macrophages (arrows) may be inserted into the sinus lumen.

The *cord of Billroth* is supported by the wing-like processes of *reticular cells* (Rt) whose surface is characteristically smooth. The spaces of the reticular cell framework contain neutrophils (N), macrophages (M) and blood platelets (P). Red blood cells and many other free cells, which must have filled the spaces in the living state, have been washed away during perfusion.

X 4,400

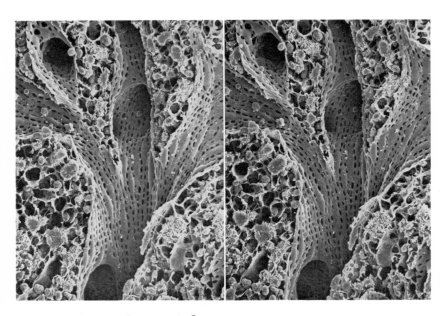

Fig. II-3 Red Pulp of the Rat in Stereo.
Branching sinuses are longitudinally opened and cordal sponge surrounds them. In the rat and in some other mammalian species the lattice structure of the sinus wall is much clearer than in human. Macrophages in the cord are large and conspicuous.

X 330

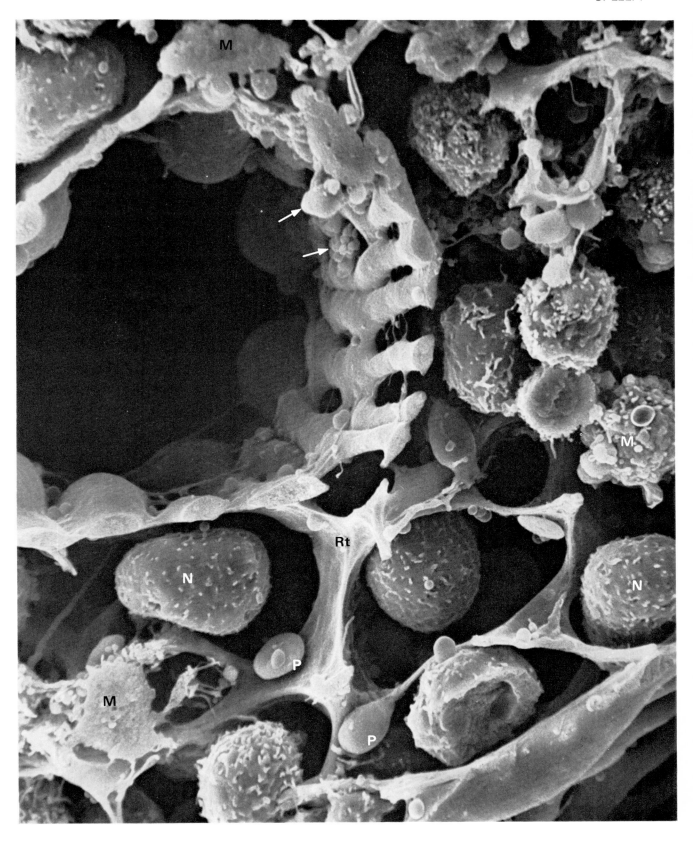

Plate II-11 Luminal Aspect of Sinus Wall. Human Spleen.

This micrograph shows the architecture of the sinus wall as seen from the luminal side. *Rod cells* run in parallel and are intermittently connected with each other by *side processes.* A nuclear swelling is shown on the bottom right. The tapered ends of a few of the rod cells are seen. The rod cells are provided with a few *thread-like microprojections* whose significance is unknown.

Parts of the cords of Billroth are shown at the top left and bottom left.

M: macrophage, N: neutrophil, L: lymphocyte.

X 5,300

(Reproduced from T. Fujita: *Arch. histol. jap.* **37**: 187—216, 1974).

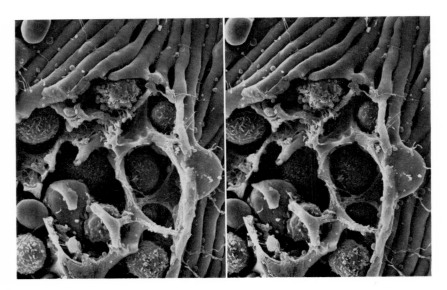

Fig. II-4 Sinus and Cord in Stereo.
A portion of tissue partly overlapping the micrograph on the opposite page is demonstrated.

A nuclear swelling of a rod cell is seen on the right hand. The extensions of reticular cells in the cord and their relation to macrophages and leukocytes may be three-dimensionally visualized.

X 1,350

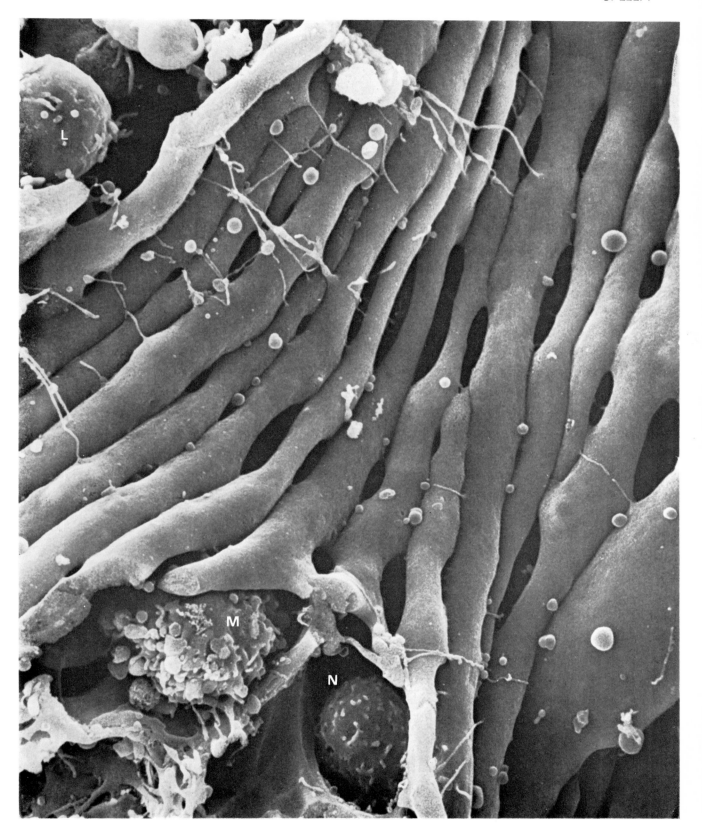

Plate II-12   Cordal Aspect of Sinus Wall.   Human Spleen.

This micrograph demonstrates a sinus seen from the outside or cordal side. At each level of the side processes of the rod cells, the cells are, on the outside, bound by the cytoplasmic processes of reticular cells. These processes are provided by cordal *reticular cells* (Rt) lying near the sinus and embrace the sinus rods in gear with transverse grooves on the outer aspect of the rods. *Ring fibers* (reticular fibers) are sandwiched between the rod cells and the reticular cell processes but they generally are not visible, hidden by the latter.

On the left, a juxta-terminal portion of an *penicillar artery* is cut crosswise. One may see a few *perforations* in the endothelium of this vessel through which blood cells may pass into the cordal space (arrows).

R: red blood cells apparently passing the sinus wall, M and m: macrophages and their processes, P: blood platelets.

X4,400

(Reproduced from T. Fujita: *Arch. histol. jap.* **37**: 187—216, 1974).

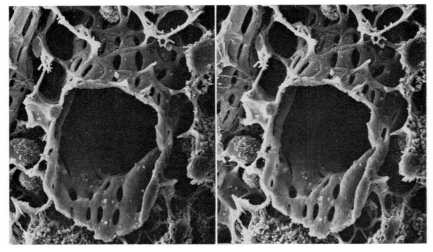

Fig. II-5 Sinus and Cord of the Rat in Stereo.
An obliquely cut sinus shows its luminal and external aspects. In the upper right a sinus wall is also viewed from the cordal side. The reticular cell feet binding the sinus rod cells are three-dimensionally demonstrated.

X 1,300

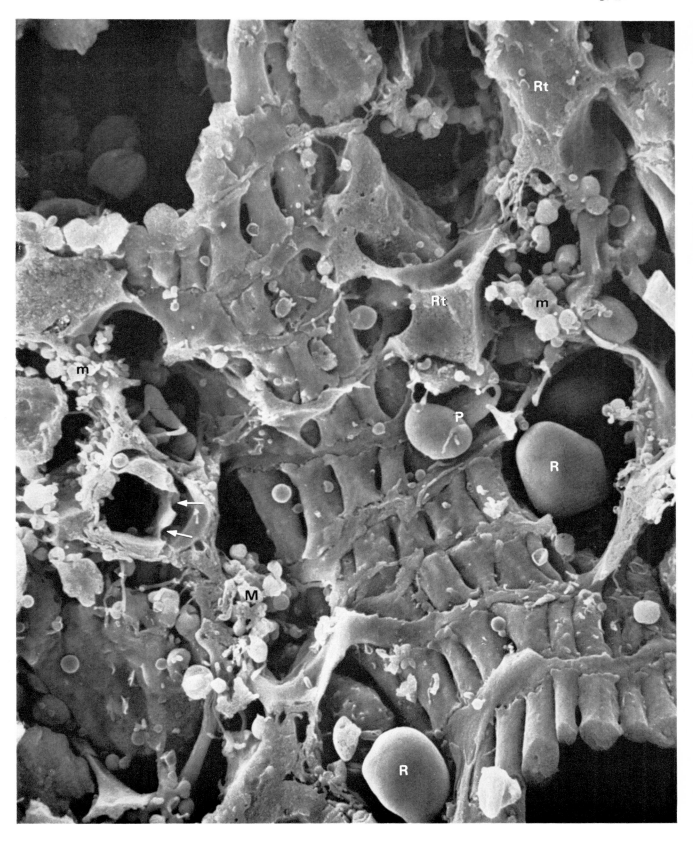

Plate II-13
A. Sheathed Artery. Human.

The branching arterioles in the red pulp are *penicillar arteries.* Before their terminal portion, they have a segment called *sheathed artery.* Here the endothelium (E) is unusually thickened and it is surrounded by concentrically flattened reticular cells (R) and macrophages (M). Reticular fibers support this sheath but they are difficult to differentiate from the fine processes of macrophages under the SEM.

    S: sinus.

    X 4,000

B. Arterial Termination. Human Spleen.

The end branches of the penicillar arteries may approach quite closely to the sinus, but they do *not* come into a direct connection with it. The endothelium of the arterial ends are either gradually perforated and fan out into the reticular cell network, or form a more or less swollen end with endothelial perforations. In both cases blood cells in the arteries are pushed into the cordal space and then, through the perforations in the sinus wall, into the sinuses.

    This micrograph shows one of such terminal swellings of penicillar arteries opened crossly. Many red blood cells apparently are passing through the endothelium. They are strongly constricted as the endothelial perforations are conspicuously small (1—2 μm as measured in other micrographs).

    S: sinus.

    X 4,600

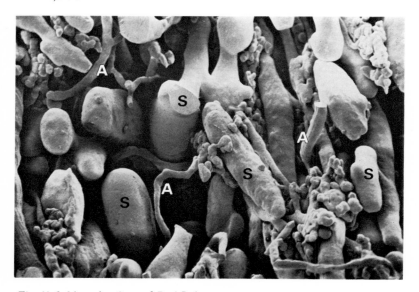

Fig. II-6 Vascular Cast of Red Pulp. Human Spleen.
This resin vascular cast of human red pulp evidences that the arterial terminals are open to the splenic cords and *not* connected directly to sinuses (open theory, see page 64). The twigs of *penicillar arteries* (A) break up into granular resin masses which correspond to the spaces in the splenic cords. The granular masses, in their turn, are attached by the sausage-like casts of *sinuses* (S).

    The cast was made by simultaneous infusion of resin from the splenic artery and vein.

    X 230

(Courtesy of Dr. T. Murakami, Department of Anatomy, Okayama University Medical School)

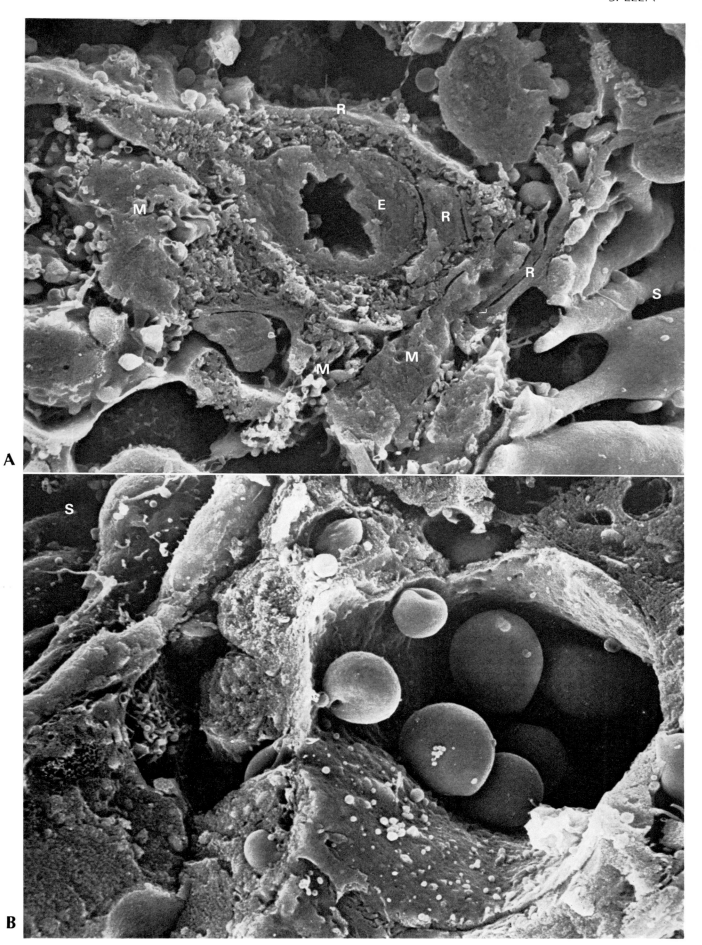

Plate II-14
A. Erythrocyte Hanging on Sinus Wall. Human.

Erythrocytes are often fixed in their hanging state on the sinus wall as seen in this micrograph. The hanging cells always protrude their heads into the sinus lumen while leaving the constricted part on the cordal side, and this indicates the passage of cells from the cord into the sinus. The *open theory* of the spleen (page 64) is thus supported, because the cells would hang on the sinus in the reverse direction, if the arterial blood would directly flow into the sinus and then come out to the cord.
   X 13,000

B. Erythrocytes and Platelets on Sinus Wall. Rat.

This micrograph shows some erythrocytes passing the perforations of the sinus wall. They must constrict their body more or less strongly. Increased rigidity of red cells inhibits this passage and the cells thus arrested may presumably be detected by patroling macrophages to be eliminated.
   P: blood platelets
   X 9,300

Fig. II-7 Blood Cells Passing Sinus Wall in Stereo. Rat.
A red cell is hanging on the sinus in the upper right, and three leukocytes protrude their body into the sinus lumen. A small part of a cordal macrophage is seen at the top.
   X 4,600

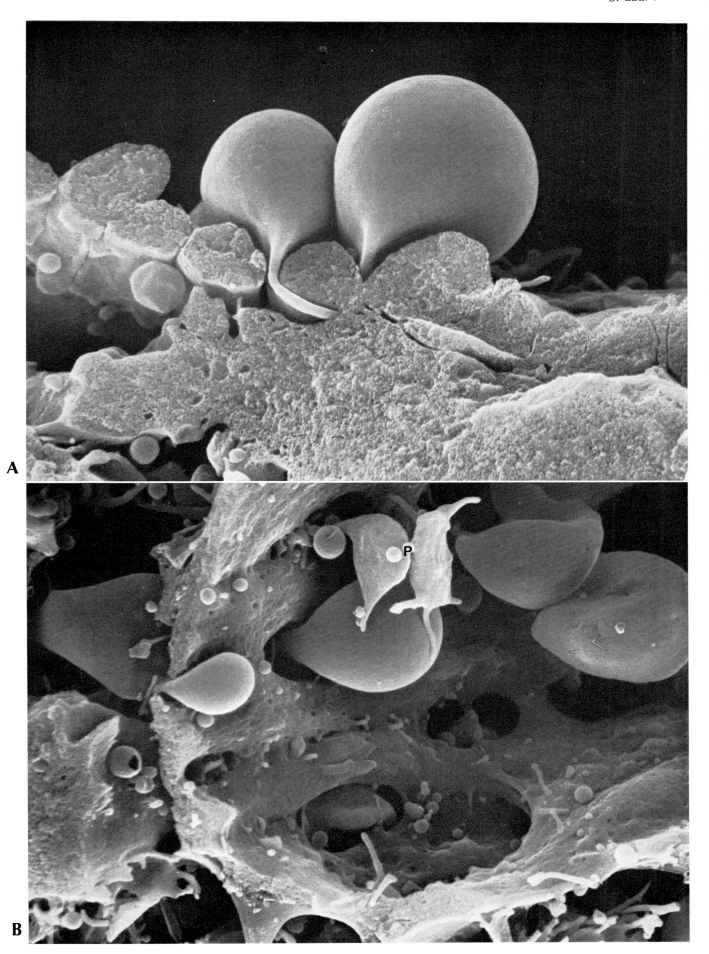

Lymph Node

Plate II-15   Lymph Node Sinus.   Mesenteric Node of the Dog.

The lymph node is a typical reticular tissue with ample spaces for lymph, but few satisfactory SEM images have been published because of the difficulty in eliminating the lymph plasm. This difficulty may be overcome by vigorous perfusion (Plate II-15, 16B) from the artery or by direct needle perfusion into the parenchyme (Plate II-16A).

This micrograph shows a *medullary sinus* of a dog lymph node spanned by the slender cytoplasmic processes of *reticular cells* (Rt). The reticular *fibers* supporting these cells are covered by them and may be exposed in fractured sites. A groove in the reticular cell which contained a reticular fiber is indicated by an arrow. Free or "round" cells, mostly *lymphocytes* (L) and *macrophages* (M) are attached to the reticular cells. The lymph coming into the node and passing through the sinus must be confronted by these phagocytotically and immunologically active cells.

In the upper left and lower right, portions of the *pulp* are shown as a more compact tissue. The round cells in this tissue are mostly lymphocytes.

The sinus is separated from the pulp by a layer of *lining cells* which are attenuated reticular cells and represent the endothelial coverage of this lymph route. *Perforations* in this cell sheet which may be the sites of free cell emigration from the pulp are obvious in the micrograph.

A: arteries in the pulp.

X 1,000

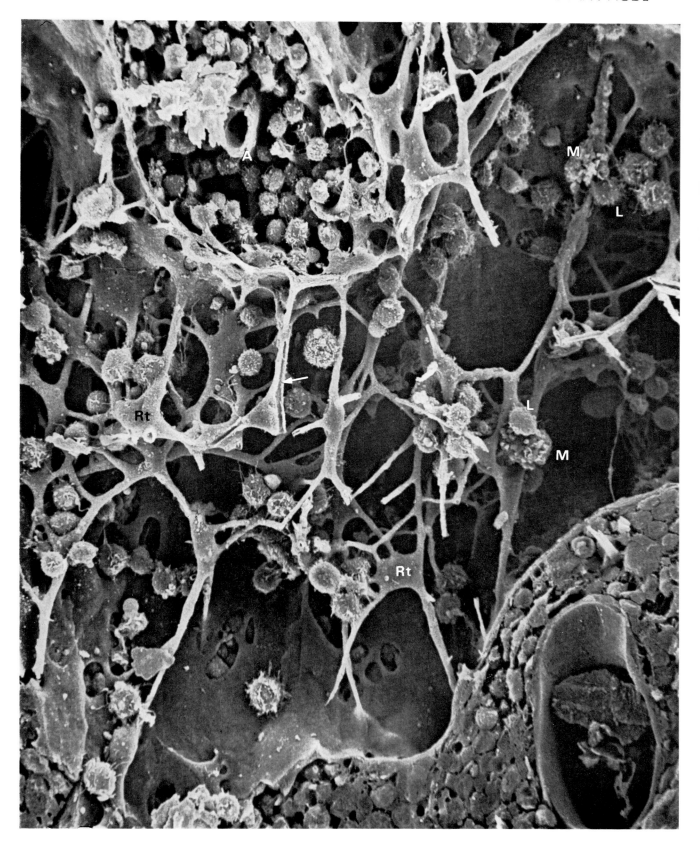

Plate II-16
A. Sinus and Pulp in Closer View.   Axillary Node of the Guinea Pig.

Reticular cells (Rt) in the sinus are smooth surfaced, though in this micrograph they are contaminated by a small amount of coagulated lymph plasm. *Macrophages* (M) in the guinea pig are especially large and, as in other species, characteristically rough surfaced. Together with Luk *et al.* (1973) we are of the opinion that the reticular cells are a type of fibroblast and *no gradations* occur between these and macrophages which belong to a specific, phagocytotic system (Fujita *et al.*, 1972).

The *pulp,* upper right in the micrograph, is bounded by a *lining cell* layer next to the sinus. The pulp consists of round cells and reticular cells which are smaller in size than the sinal reticular cells.

   X 1,900

B. Trabecule of Lymph Node.   Mesenteric Node of the Dog.

The lymph node tissue is supported by the capsule and trabecules composed of thick collagen fibers. This micrograph shows a trabecule revealing its blood vessels and collagen bundles transversely cut. The latter fan out into fine fibrils supporting the parenchyme of the node. Facing the sinus, the surfaces of the trabecule and its branches are covered by a thin layer of lining and reticular cells.

   X 780

(Plate II-16A: Courtesy of Dr. M. Muto, Department of Anatomy, Niigata University School of Medicine)

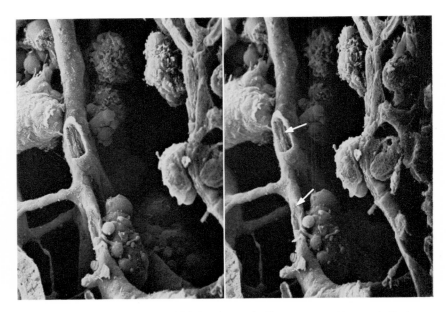

Fig. II-8 Sinus Reticulum and Lining Layer in Stereo.   Dog Mesenteric Node.
In this reticulum of sinus, collagen fibrils (reticular or argyrophil fibers) covered by a reticular cell sheet are exposed by fracture (arrows). On the right hand a lining cell layer and pulp tissue are demonstrated.
   X 1,500

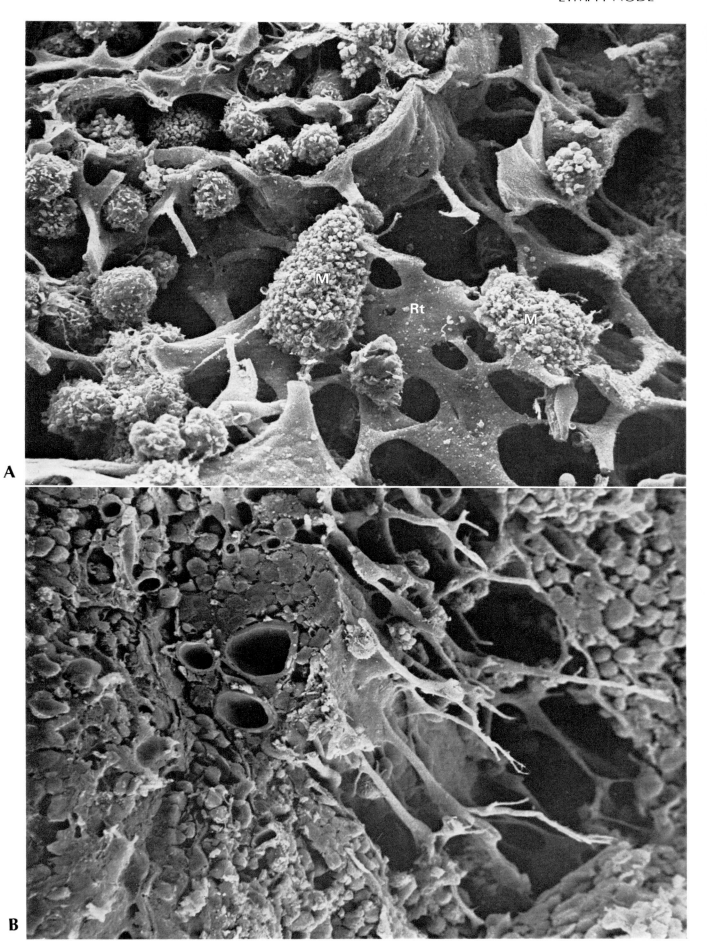

Tonsil

Plate II-17   Cryptal Epithelium of Tonsil.   Human Palatine Tonsil.

The tonsil is a lymphoid organ covered by a stratified squamous epithelium which is folded to form deep crypts. These micrographs show surface views of the epithelium lining a crypt of a palatine tonsil surgically removed from an 8-year-old boy. The *cryptal epithelium* shows here and there an *oval hole* filled with rounded cells which were called "fungiform cells" by Oláh and Everett (1975) in their SEM study of the rabbit palatine tonsil. These authors interpret that the fungiform cells are a type of epithelial cells, but their true nature deserves further investigation.

A is an overview of a *fungiform cell patch.* Careful observation of this picture suggests that there may be transitional forms between the fungiform cells and the cells forming the stratified epithelium. Some of the fungiform cells are covered by microvilli. A few small round cells(∗) may be *lymphocytes* passing through the epithelium.

It seems an attractive hypothesis that the fungiform cell patch represents a pass through which antigenic and invasive information from the oral cavity is introduced into the lymphoid tissue and antibodies, lymphocytes and other protective elements may come out.

B shows another part of the cryptal epithelium. The squamous epithelial cells are noteworthily polymorphous. The *microplicae* conspicuously differ in density from cell to cell; a few cells are covered by *microvilli* instead. A peculiar, long projection of an apparently waste cell is shown in the middle. At the bottom are two fungiform cell patches.

C is a closer view of one of the patches. Besides fungiform cells covered by microvilli, a cell possessing *cilia,* besides microvilli, is demonstrated. Bodies labeled ∗ possibly represent the heads of migrating lymphocytes.

A: × 5,900,   B: × 2,100,   C: × 10,000

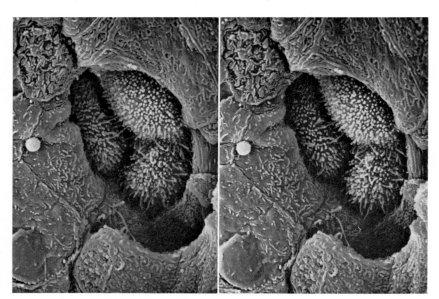

Fig. II-9 Fungiform Cell Patch in Stereo. Human Tonsil.
This stereo pair demonstrates that the patch may be quite a deep hole and fungiform cells lie as if to plug its bottom.

× 3,700

(Plate II-17A: Produced by Dr. Y. Umetani, Department of Otolaryngology, Niigata University School of Medicine).

(All other micrographs of this plate and Fig. II-9 were taken also from the specimens provided by Dr. Umetani).

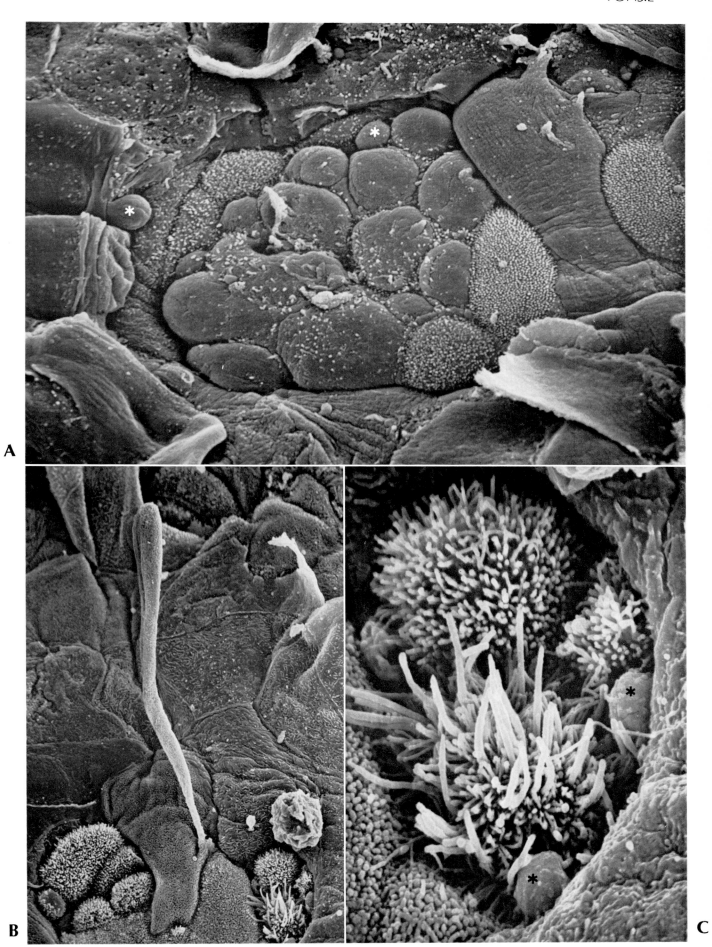

Plate II-18
A. Postcapillary Venule. Rabbit Tonsil.

The postcapillary venule is found in secondary lymphoid organs like the lymph node and tonsil and characterized by a peculiar, *high endothelial* lining. Lymphocytes freed from the thymus and recirculating in the blood, are called T cells, and are known to pass selectively through this specialized epithelium to reside in the area of the lymphoid organ surrounding the postcapillary venule (thymus dependent area).

The real shape of and connections between the endothelial cells of this venule had been obscure in spite of numerous TEM studies, but became clear from recent SEM observations (Umetani, 1977; Cho and De Bruyn, 1979). The endothelium characteristically consists of *stellate cells* with a swollen body and several *side processes* which are interwoven with the processes of adjacent cells.

X 3,200

B. T Cell and the Venule Endothelium. Rabbit Tonsil.

This micrograph is a closer view of the postcapillary venule surface. The endothelial cells are interdigitated with mighty *side processes*. Granular and villous microprojections may occur on the cell surface. Thin folds may mark the cell boundaries (arrows).

A *lymphocyte* of probable T-type (because of its smooth surface with few projections) is seen under migration through the venule wall, apparently pushing its head into the intercellular space of the endothelium.

X 18,500

(Plate II-18: Courtesy of Dr. Y. Umetani, Department of Otolaryngology, Niigata University School of Medicine).

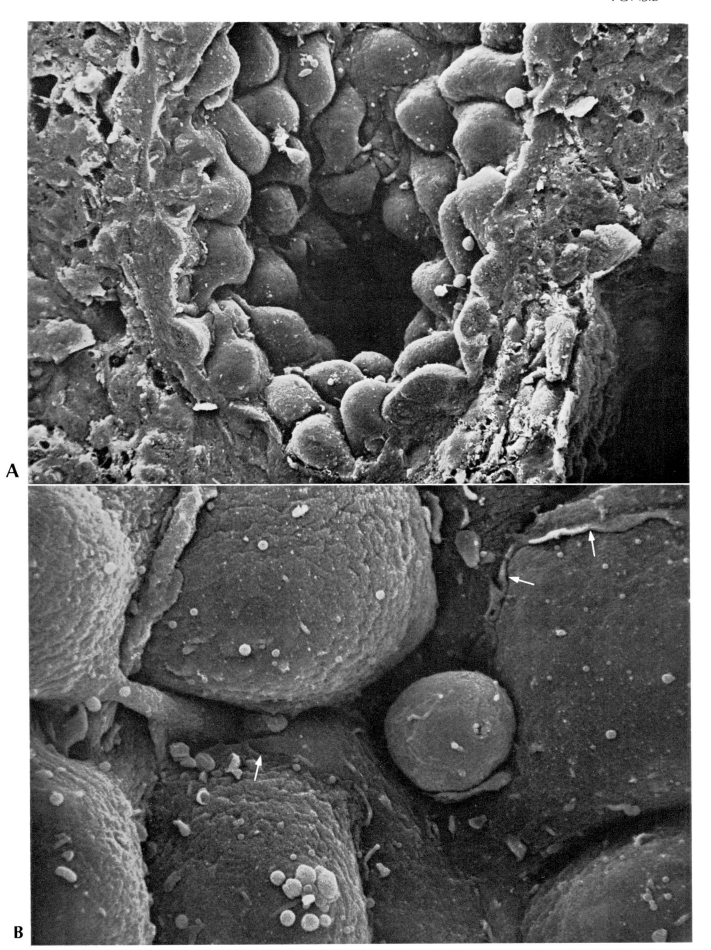

A

B

Blood Cells

Human *red blood cell* represented one of the earliest biological materials successfully observed by SEM (Clarke and Salsbury, 1967). Numerous papers have then reported different pathological changes of erythrocytes as well as their artifactual deformations (Tokunaga *et al.*, 1969; Bessis *et al.*, 1973). Researchers' attention meanwhile was paid to the surface structure of white cells or *leukocytes*. Some researchers succeeded to identify the main cell types under the SEM by directly comparing them with their stained light microscopic images (Hattori, 1972), while others focussed their attention on the structure of migrating leukocytes (Felix *et al.*, 1978).

Lymphocytes have been studied most extensively under the SEM and this volume also contains figures of lymphocytes in various attitudes such as being packed in a lymph node pulp (Plate II-15), attached to blood vascular endothelium (Plate II-7), or conjugated with a macrophage (Fig. I-1). The exciting reports by Lin *et al.* (1973 a, b) that the *B cells* are generally covered much more densely with microvilli than *T cells* have stimulated numerous immunologists towards SEM analysis of lymphocyte types. Yet it has been noticed that the above criteria of B and T cells are applicable only to a majority of lymphocytes and a considerable number of them show atypical or even reverse morphology of cell processes (Polliack *et al.*, 1974).

Blood platelets are one of the best objects of high-resolution SEM. Their subtle metamorphosis due to mechanical and chemical stimuli would be studied more effectively by none of other methodologies than by SEM (Hattori *et al.*, 1969, 1978). Both *in vivo* and *in vitro* studies, using SEM, are actively exploring the mechanisms and pathological changes in the hemostatic process and thrombus formation in which blood platelets are primarily involved.

Plate II-19

A. Red Blood Cells.   From Healthy Human.

Erythrocytes of human and most mammalian species are *biconcave disks*. They are highly flexible and may undergo conspicuous deformation when squeezed through a narrow space as seen in Plates II-13B and 14.

X 7,800

B. Red Cells of Spherocytosis.   In Patient's Spleen.

Spherocytosis is a form of anemia characterized by occurrence of irregular-shaped red cells. Spheroid cells may be numerous but may occupy only a small part of erythrocytes. The only known treatment of this anemia is splenectomy, which was performed also in this patient, a 7-year-old girl.

This micrograph shows a fracture view of Billroth's cord of the spleen removed. The cordal spaces store an unusually large number of irregular-shaped red cells, as these deformed cells can not smoothly pass the sinus wall of the spleen.

The spheroid and rounded-polyhedral red cells possess variable numbers of needle-point *pits* which have been first recognized in spherocytes by Salsbury and Clarke (1967).

Platelets (P) are seen on the reticular cells of the splenic cord.

X 6,200

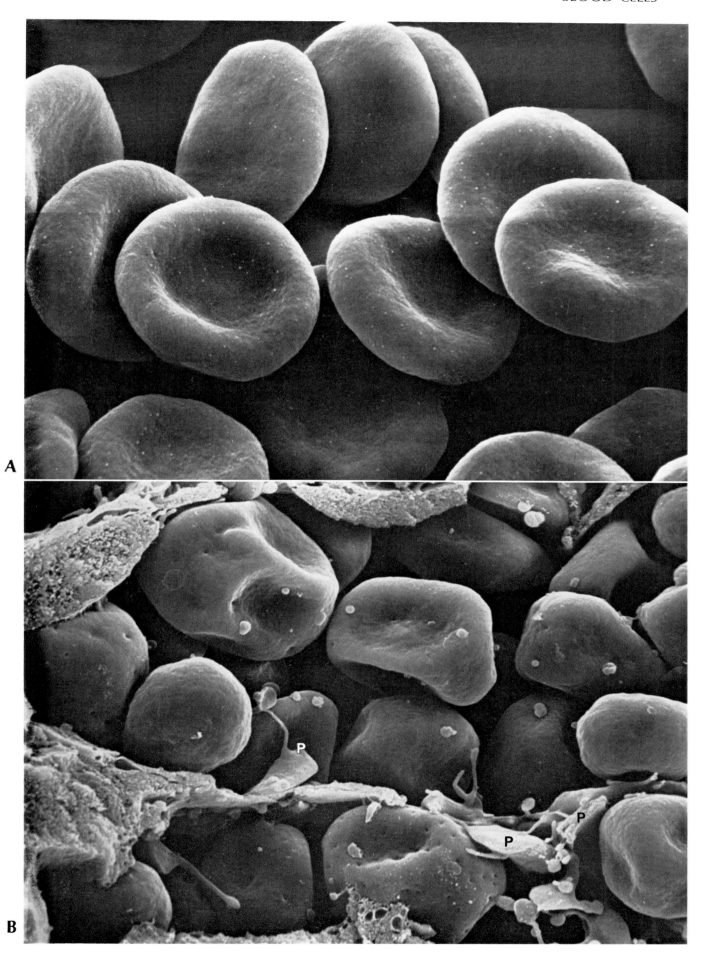

Plate II-20

A. Neutrophil Leukocytes. From Healthy Human Blood.

Neutrophils in their rounded, circulating form are characterized by short and twisted microvilli which are partly lying on the cell surface, forming microplicae, and only their rounded end stands out. There are mouth-like indentations which may be accentuated by surrounding microplicae.

P: platelets, R: red blood cell.

X 16,000

B. Moving Leukocyte. From Healthy Human Blood.

A probable neutrophil under ameboid locomotion on the glass surface of a hemocytometer is shown in this micrograph. The cell is moving in the direction indicated by the arrow. Ahead is a tongue-like *pseudopodium* covered by lamellar microprojections, which are known by light microscope observation of living cells to undergo an active ruffling movement. The hind portion of the cell usually is tapered and stumped. This tail often projects several spiny *filopodia* as well as irregular microprojections.

P: platelet

X 6,500

(Plate II-20: Courtesy of Dr. A. Hattori, First Department of Internal Medicine, Niigata University School of Medicine).

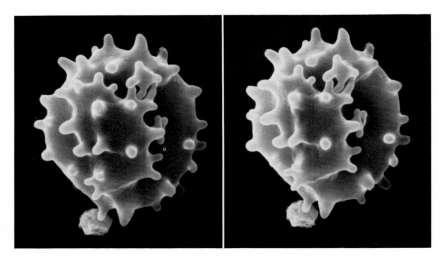

Fig. II-10 Echinocyte (Deformed Erythrocyte) in Stereo.
Red blood cells may transform into cells called echinocytes (echinos: a prickly husk; hedgehog) during manipulation or preservation of blood samples. This deformation is favored by a lowered membrane resistance of red cells which may occur in certain pathological cases, including spherocytosis above mentioned.

This micrograph shows an echinocyte found in a blood specimen derived from a nephrotic child and prepared by normal procedures. The form, like this cell, with papillar projections may be called a sphero-echinocyte, while a form with stronger specules, a drepano-echinocyte (Bessis, 1973; Bessis *et al.*, 1973).

X 9,000

(The specimen was provided by Prof. K. Sakai, Department of Pediatrics, Niigata University School of Medicine).

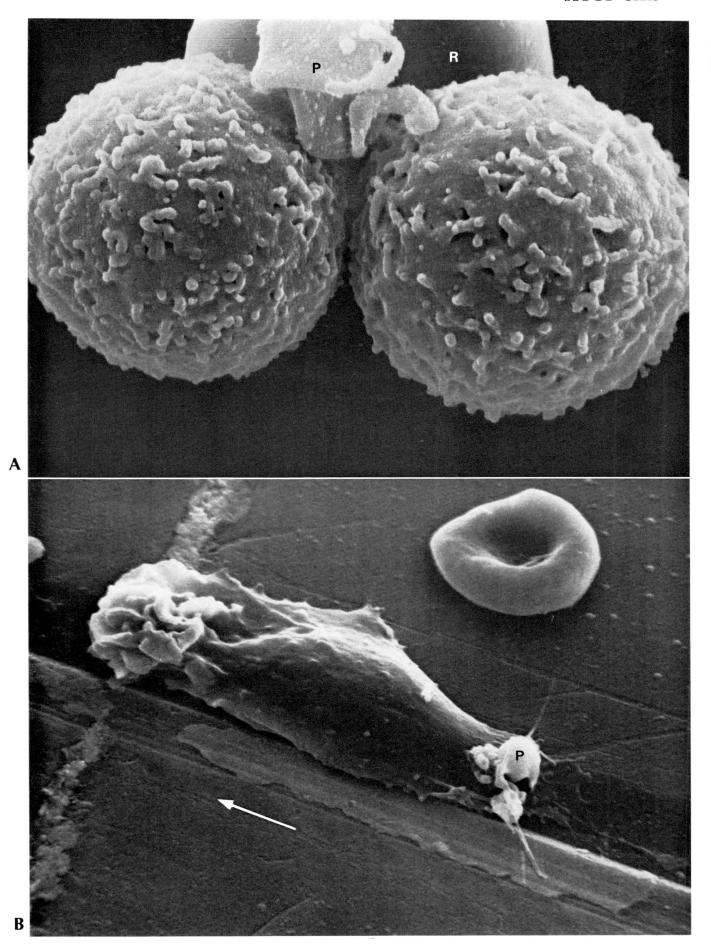

Plate II-21
A. Lymphocyte : B Cell. From Healthy Human Blood.

The B or bone marrow-derived type of lymphocytes generally is characterized by *microvilli*, not very long, fairly uniform in thickness and partly twisted, which densely cover the cell surface.
   X 18,000

B. Lymphocyte : T Cell. From Healthy Human Blood.

The T or thymus-derived type of lymphocytes generally is smooth in surface possessing only a few microprojections. The microprojections may be irregular in shape as in this micrograph but may also be represented by long, tentacle-like microvilli or filopodia.
   X 16,000

C. Monocyte with Pseudopodium. From Healthy Human Blood.

Monocytes have a rather irregular surface structure possessing villous, lamellar and granular microprojections. This micrograph shows a monocyte which projects on a glass surface, a pseudopodium for migration (right hand). Note that the pseudopodium issues long microvilli.
   X 19,000

(Plate II-21: Courtesy of Dr. A. Hattori, First Department of Internal Medicine, Niigata University School of Medicine).

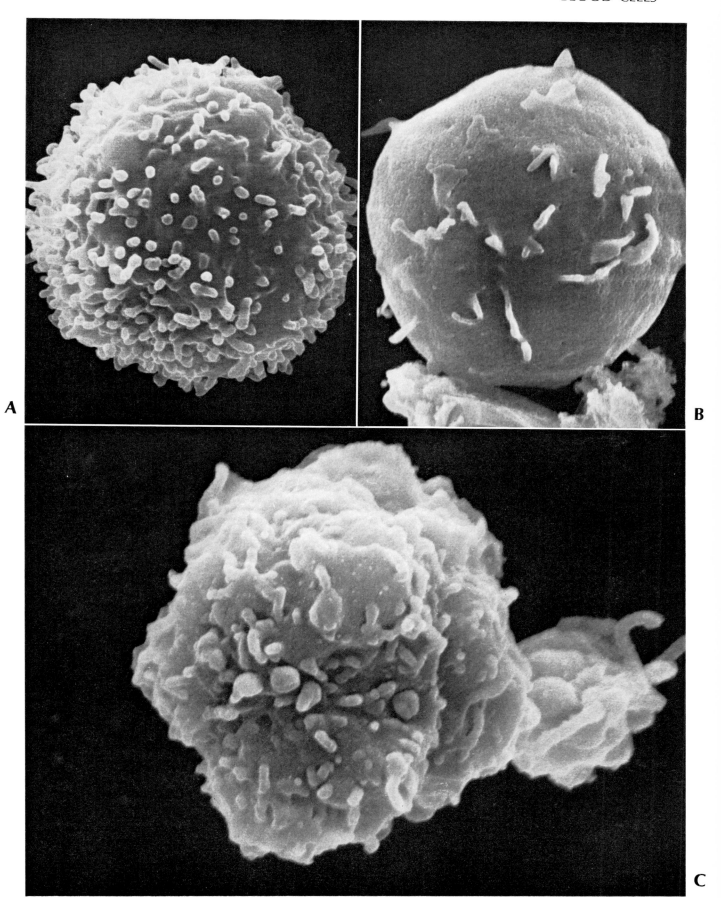

Megakaryocyte and Blood Platelets

Blood platelets are formed by fragmentation of the cytoplasm of megakaryocytes. In man megakaryocytes usually occur only in bone marrow, while some mammals contain them also in the spleen. The splenic megakaryocytes of the mouse have been a favored material for the study of platelet formation.

Plate II-22
A. Megakaryocyte in the Splenic Cord. Mouse.

Megakaryocyte is a large cell, over 25µm in diameter, showing ovoid or irregular shapes. The macrophage shown in this micrograph reveals its characteristic surface provided with microvilli and pores. A cytoplasmic branch is fractured on the upper part of the cell (*), so that the interior structure, especially the *cleft system* of the cytoplasm is obvious.
   X 8,000

B, C. Cytoplasm of Platelet Forming Megakaryocyte. Mouse Spleen.

Both micrographs show how megakaryocytes are fragmented into blood platelets. Tortuous channels and clefts are formed in the intermediate zone of cytoplasm which comprise a *platelet demarkating membrane system* (DMS). TEM studies (Yamada, 1957; Behnke, 1968; etc.) have shown that the DMS is continuous to the plasma membrane through the cell surface pores (arrows) and many authors believe that the DMS is formed by invagination of plasma membrane (Behnke, 1968; MacPherson, 1972).

   In Figure B, small territories for individual platelets are being demarkated. Figure C shows a more advanced stage, and irregular-shaped platelets are almost completely separated. In both pictures a layer of ectoplasm (E) remains unseparated. Platelets are generally pushed out from some loose sites occurring in the ectoplasm; the fate of the ectoplasm is controversial (Ihzumi *et al.*, 1977).
   G: Golgi area
   B: X 23,000,   C: X 23,000

(Plate II-22A-C: Courtesy of Dr. T. Ihzumi, First Department of Internal Medicine, Niigata University School of Medicine).
(Plate II-22C: From Ihzumi, T., A. Hattori, M. Sanada and M. Muto: *Arch. histol. jap.* 40: 305—320, 1977).

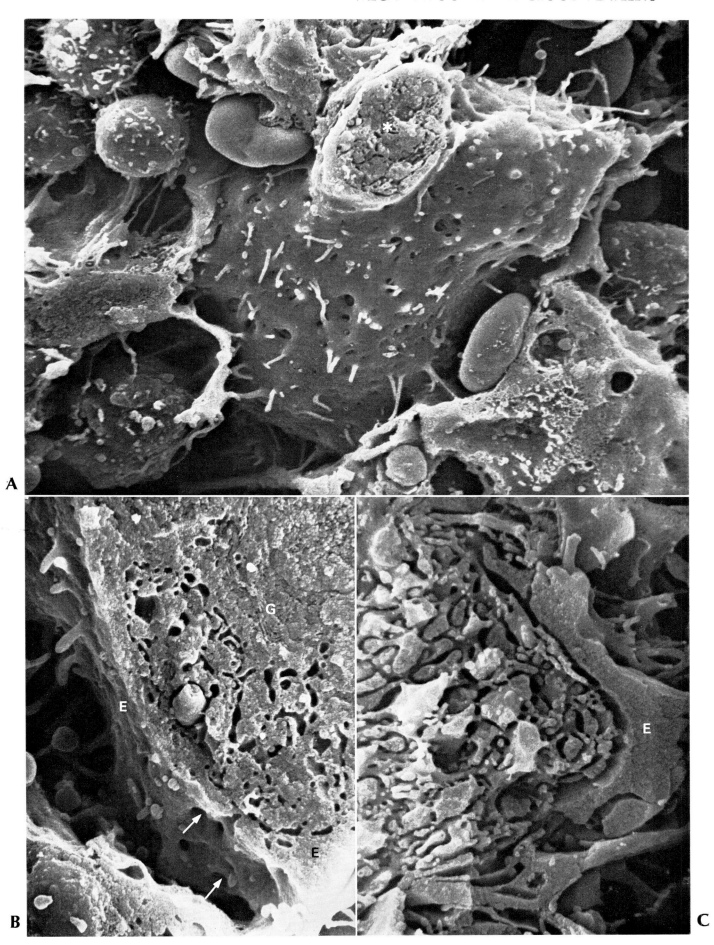

Plate II-23
A. Megakaryocyte Fragmented into Platelets. Mouse Spleen.

In this magakaryocyte, the cytoplasm, apparently including its superficial layer, is entirely fragmented into blood platelets, which partly have initiated their migration. An ectoplasmic layer is not identifiable. It has been demonstrated by a study in mouse spleen that, at least in some megakaryocytes, ectoplasm becomes fragmented giving rise to so-to-say false platelets devoid of organelles characteristic to ordinary platelets (Ihzumi *et al.*, 1977).

   X 6,500

B. Platelet Ribbons Projected by Megakaryocyte. Rat Bone Marrow.

Megakaryocyte may be fragmented into platelets in the extrasinal reticular tissue but, as shown in the early light microscope study by Wright (1910), also may project its processes ("pseudopods") into sinus lumen, which then become divided into platelets. In the rat bone marrow, such magakaryocyte processes are found numerously within the sinus and represented, characteristically for this species, by ribbons of 1—2 μm thickness. Occurrence of beaded and consticted shapes of these ribbons together with separated platelets (P) indicates the stages of their *segmentation* into individual platelets.

   In contrast to red and white blood cells migrating through intracellular fenestrations of the sinus endothelium, platelets and their ribbons usually come into blood stream through intercellular spaces.

   X 4,500

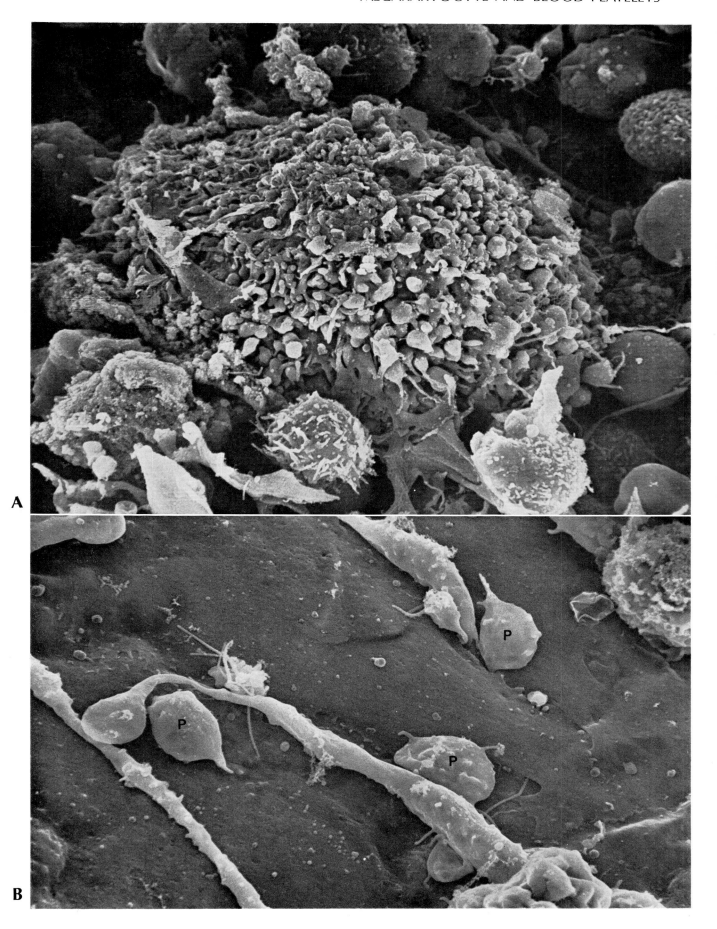

Plate II-24
A. Blood Platelets. From Human Blood.

Blood platelets play important roles in hemostasis by stimulating the process of blood clotting and by forming thrombi to plug deficiencies in the vessel wall. In normal, unstimulated condition in the blood, they are mostly lens-like disks 2—4 μm in diameter. Forms conspicuously deviated from the discoid shape, like the *club-shaped* one shown here have been found in about 4% of platelets sampled from numerous healthy subjects (Hattori *et al.*, 1977).

   X 15,500

B. Platelets Activated. Human.

This micrograph shows human platelets which have been slightly activated during the blood sampling. They project microvillous or, better to say, filopodial microprojactions (the designation "pseudopodia" widely accepted by hematologists seems inappropriat: see page 11) from the margin of the disk.

   X 15,500

C. Collagen Induced Platelet Aggregate. Human.

When the platelets are stimulated, as in this case, by collagen they are swollen and project, besides the filopodia, clumpy projections. These platelets tend to gather forming a large aggregate. Arrows indicate *fibrin fibers* whose formation is mediated by thromboplastin produced by the platelets.

   X 4,000

(Plate II-24: Courtesy of Dr. A. Hattori, First Department of Internal Medicine, Niigata University School of Medicine).

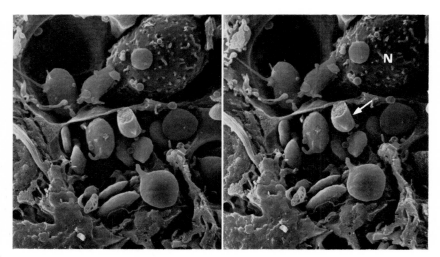

Fig. II-11 Splenic Platelets in Stereo. Human.
Platelets are very numerous in the spleen, especially in the cord of Billroth. In the spaces bounded by attenuated processes of reticular cells, discoid platelets are seen partly projecting filopodial and shorter microprojections. The arrow shows a fractured platelet.
   N: neutrophil leukocyte.
   X 3,000

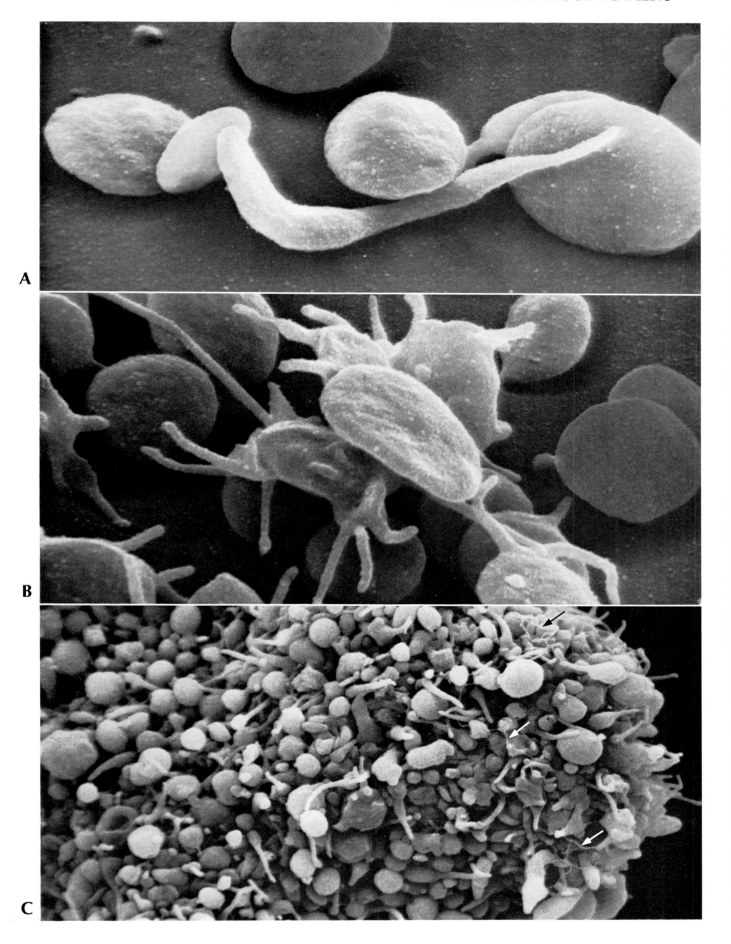

Plate II-25　Blood Platelets in Early Hemostatic Process.　Rabbit.

In the rabbit carotid artery, the endothelium was mechanically removed and the hemostatic process occurring there was observed (Hattori *et al.*, 1975).

A shows the *adhesion* of platelets to the damaged vessel wall (1 sec after the damage). Several platelets projecting filopodial and angular microprojections from their irregular margin are attached to the exposed subendothelial collagen fibrils. The microprojections apparently play the principal role in the platelet recognition of tissue damage and adhesion to the site.

B demonstrates the starting *aggregation* of platelets which adhere more and more numerously on the exposed tissue (15 sec after the damage). Some platelets are flatly extended as if to cover as large damage area as possible, while others are swollen and rounded. A few platelets cover others which adhered earlier, thus forming layered aggregates.

C shows the surface view of a hemostatic plug which is formed by tightly aggregated and thickly piled up platelets (15 sec after injury). The platelets are markedly variable in shape. Long filopodia tend to be extended in the direction of blood stream.

A: X 12,000,　B: X 11,000,　C: X 9,200

(Plate II-25: Courtesy of Dr. A. Hattori, First Department of Internal Medicine, Niigata University School of Medicine).

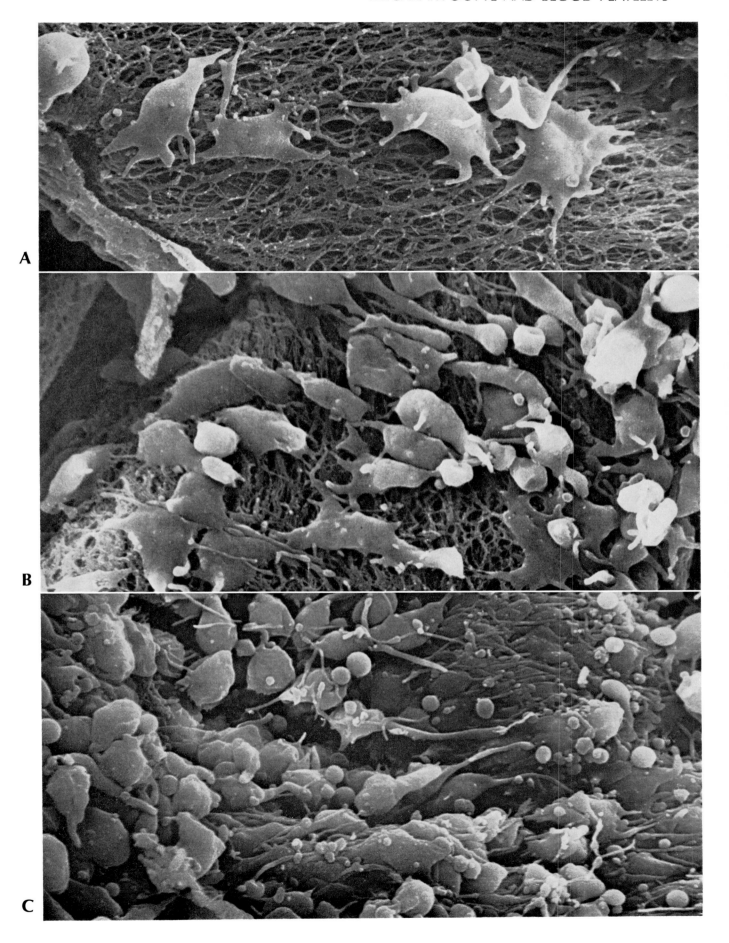

References

Albert, E. N. and R. K. Nayak: Surface morphology of human aorta as revealed by the scanning electron microscope. *Anat. Rec.* **185**: 223—234 (1976).

Becker, R. P. and P. P. H. De Bruyn: The transmural passage of blood cells into myeloid sinusoids and the entry of platelets into the sinusoidal circulation; a scanning electron microscopic investigation. *Amer. J. Anat.* **145**: 183—206 (1976).

Behnke, O.: An electron microscope study of the megakaryocyte of the rat bone marrow. I. The development of the demarcation membrane system and the platelet surface coat. *J. Ultrastr. Res.* **24**: 412—433 (1968).

Bessis, M.: Living blood cells and their ultrastructure. Springer, Berlin—Heidelberg—New York, 1973.

Bessis, M., R. I. Weed and P. F. Leblond: Red Cell Shape. Physiology, Pathology, Ultrastructure. Springer Verlag, New York—Heidelberg—Berlin, 1973.

Campbell, F. R.: Ultrastructural studies of transmural migration of blood cells in the bone marrow of rats, mice and guinea pigs. *Amer J. Anat.* **135**: 521—536 (1972).

Cho, Y. and P.H. De Bruyn: The endotherial structure of the postcapillary venules of the lymph node and the passage of lymphocytes across the venule wall. *J. Ultrastr. Res.* **69**: 13—21 (1979).

Clarke, J. A. and J. Salsbury: Surface ultrastructure of human blood cells. *Nature* **215**: 402—404 (1967).

De Bruyn, P. P. H. and Y. Cho: Contractile structures in endothelial cells of splenic sinusoids. *J. Ultrastr. Res.* **49**: 24—33 (1974).

De Bruyn, P. P. H., S. Michelson and R. P. Becker: Endocytosis, transfer tubules, and lysosomal activity in myeloid sinusoidal endothelium. *J. Ultrastr. Res.* **53**: 133—151 (1975).

Edanaga, M.: A scanning electron microscope study on the endothelium of the vessels. I. Fine structure of the endothelial surface of aorta and some other arteries in normal rabbits. *Arch. histol jap.* **37**: 1—14 (1974).

Edanaga, M.: A scanning electron microscope study on the endothelium of vessels. II. Fine surface structure of the endocardium in normal rabbits and rats. *Arch. histol. jap.* **37**: 301—312 (1975)

Evan, A. P., W. G. Dail, D. Dammrose and C. Palmer: Scanning electron microscopy of cell surfaces following removal of extracellular material. *Anat. Rec.* **185**: 433—446 (1976).

Fawcett, D. W.: Comparative observations on the fine structure of blood capillaries. In: (ed. by) J. L. Orbison & D. Smith: The Peripheral Blood Vessels. Williams and Wilkins, Baltimore, 1963. (p. 17—44).

Felix, H., G. Haemmerli and P. Sträuli: Dynamic Morphology of Leukemia Cells. A Comparative Study by Scanning Electron Microscopy and Microcinematography. Springer Verlag, Berlin—Heidelberg—New York, 1978.

Fujita, T.: A scanning electron microscope study of the human spleen. *Arch. histol. jap.* **37**: 187—216 (1974).

Fujita, T., M. Miyoshi and T. Murakami: Scanning electron microscope observation of the dog mesenteric lymph node. *Z. Zellforsch.* **133**: 147—162 (1972).

Fujita, T., J. Tokunaga and M. Edanaga: Scanning electron microscopy of the glomerular filtration membrane in the rat kidney. *Cell Tiss. Res.* **166**: 299—314 (1976).

Fukuta, K., T. Nishida and K. Mochizuki: Electron microscopy of the splenic circulation in the chicken. *Jap. J. vet. Sci.* **38**: 241—254 (1976).

Grisham, J. W., W. Nopanitaya, J. Compagno and A. E. H. Nägel: Scanning electron microscopy of normal rat liver: the surface structure of its cells and tissue components. *Amer. J. Anat.* **144**: 295—322 (1976).

Hattori, A.: Scanning electron microscopy of human peripheral blood cells. *Acta hematol. jap.* **35**: 457—482 (1972).

Hattori, A., C. Jinbo, T. Iizumi, S. Ito and M. Matsuoka: A scanning electron microscope study on hemostatic reaction. Early hemostatic plug formation and the effect of aspirin. *Arch. histol. jap.* **37**: 343—364 (1975).

Hattori, A., M. Sanada, T. Iizumi, S. Ito, T. Izumi and M. Matsuoka: Study on platelet shape and its relation to function. 1st Report: A method on platelet shape and the results in normal subjects. *Blood & Vessel* **8**: 588—597 (1977).

Hattori, A., J. Tokunaga, T. Fujita and M. Matsuoka: Scanning electron microscopic observations on human blood platelets and their alterations induced by thrombin. *Arch. histol. jap.* **31**: 37—54 (1969).

Hattori, A., T. Watanabe and T. Izumi: Scanning electron microscopy study on hemostatic reaction. Mural thrombus after the removal of endothelium, with special references to platelet behavior, site of fibrin formation and microhemolysis. *Arch. histol. jap.* **41**: 205—227 (1978).

Hudson, G. and J. M. Yoffey: The passage of lymphocytes through the sinusoidal endothelium of guinea-pig bone marrow. *Proc. Roy. Soc. (B)* **165**: 486—495 (1966).

Ihzumi, T., A. Hattori, M. Sanada and M. Muto: Megakaryocyte and platelet formation: A scanning electron microscope study in mouse spleen. *Arch. histol. jap.* **40**: 305—320 (1977).

Irino, S., T. Murakami and T. Fujita: Open circulation in the human spleen. Dissection scanning electron microscopy of conductive-stained tissue and observation of resin vascular casts. *Arch. histol. jap.* **40**: 297—304 (1977).

Korneliussen, H.: Fenestrated blood capillaries and lymphatic capillaries in rat skeletal muscle. *Cell Tiss. Res.* **163**: 169—174 (1975).

Leblond, P.-F.: Étude, au microscope électronique a balayage, de la migration des cellules sanguines a travers les parois des sinusoïdes spléniques et médullaires chez le Rat. *Nouv. Rev. Franç. Hématol.* **13**: 771—788 (1973).

Lin, P. S., A. G. Cooper and H. H. Wortis: Scanning electron microscopy of human T-cell and B-cell rosettes. *New Engl. J. Med.* **289**: 548 (1973b).

Lin, P. S., S. Tsai and D. F. H. Wallach: Major differences in the surface morphology of thymocytes and peripheral lymphocytes are revealed by scanning electron microscopy. In: (ed. by) E. Gerlach *et al.*: 2nd Int. Symp. on Metabolism and Membrane Permeability of Erythrocytes, Thrombocytes and Leukocytes. Georg Thieme, Stuttgart, 1973a (p. 438—440).

Luk, S. C., C. Nopajaroonsri and G. T. Simon: The architecture of the normal lymph node and hemolymph node. A scanning and transmission electron microscopic study. *Lab. Invest.* **29**: 258—265 (1973).

MacPherson, G. G.: Origin and development of the demarcation system in megakaryocytes of rat bone marrow. *J. Ultrastr. Res.* **40**: 167—177 (1972).

Miyoshi, M. and T. Fujita: Stereo-fine structure of the splenic red pulp. A combined scanning and transmission electron microscope study on dog and rat spleen. *Arch. histol. jap.* **33**: 225—246 (1971).

Miyoshi, M., T. Fujita and J. Tokunaga: The red pulp of the rabbit spleen studied under the scanning electron microscope. *Arch. histol. jap.* **32**: 289—306 (1970).

Motta, P.: A scanning electron microscopic study of the rat liver sinusoid. Endothelial and Kupffer cells. *Cell Tiss. Res.* **164**: 371—385 (1975).

Motta, P. and K. R. Porter: Structure of rat liver sinusoids and associated tissue spaces as revealed by scanning electron microscopy. *Cell Tiss. Res.* **148**: 111—125 (1974).

Murakami, M., A. Sugita, T. Shimada and K. Nakamura: Surface view of pericytes on the retinal capillary in rabbits revealed by scanning electron microscopy. *Arch. histol. jap.* **42**: 297—303 (1979).

Muto, M.: A scanning electron microscopic study on endothelial cells and Kupffer cells in rat liver sinusoids. *Arch. histol. jap.* **37**: 369—386 (1975).

Muto, M.: A scanning and transmission electron microscopic study on rat bone marrow sinuses and transmural migration of blood cells. *Arch. histol. jap.* **39**: 51—66 (1976).

Muto, M., M. Nishi and T. Fujita: Scanning electron microscopy of human liver sinusoids. *Arch. histol. jap.* **40**: 137—151 (1977).

Oláh, I. and N. B. Everett: Surface epithelium of the rabbit palatine tonsil: Scanning and electron microscopic study. *J. Reticuloendothel. Soc.* **18**: 53—62 (1975).

Peine, C. J. and F. N. Low: Scanning electron microscopy of cardiac endothelium ofthe dog. *Amer. J. Anat.* **142**; 137—158 (1975).

Pictet, R., L. Orci, W. G. Forssmann and L. Girardier: An electron microscope study of the perfusion-fixed spleen. I. The splenic circulation and the RES concept. *Z. Zellforsch.* **96**: 372—399 (1969).

Polliack, A., S. M. Fu, S. D. Douglas, Z. Bentwich, N. Lampen and E. de Harven: Scanning electron microscopy of human lymphocyte-sheep erythrocyte rosettes. *J. Exp. Med.* **140**: 146 (1974).

Saito, H.: Fine structure of the reticular cells in the rat spleen, with special reference to their fibro-muscular features. *Arch. histol. jap.* **40**: 333—345 (1977).

Saito, H. and T. Takagi: Differentiation of the oral lymphoid tissues: Surface structures of the tonsilar epithelium of the developing rabbit. *Acta anat. nippon.* **51**: 168—179 (1976).

Salsbury, A. J. and J. A. Clarke: New method for detecting changes in the surface apperance of human red blood cells. *J. clin. Pathol.* **20**: 603—610 (1967).

Smith, U., J. W. Ryan, D. D. Miche and D. S. Smith: Endothelial projections as revealed by scanning electron microscopy. *Science* **173**: 925—926 (1971).

Suzuki, T.: Application of scanning electron microscopy in the study of the human spleen: Three dimensional fine structure of the normal red pulp and its changes as seen in splenomegalias associated with Banti's syndrome and cirrhosis of the liver. *Acta haematol. jap.* **35**: 506—522 (1972).

Suzuki, T., M. Furusato, S. Takasaki, S. Shimizu and Y. Hataba: Stereoscopic scanning electron microscopy of the red pulp of dog spleen with special reference to the terminal structure of the cordal capillaries. *Cell Tiss. Res.* **182**: 441—453 (1977).

Tokunaga, J., T. Fujita and A. Hattori: Scanning electron microscopy of normal and pathological human erythrocytes. *Arch. histol. jap.* **31**: 21—35 (1969).

Tokunaga, J., and T. Fujita: Endothelial surface of rabbit aorta as observed by scanning electron microscopy. *Arch. histol. jap.* **36**: 129—141 (1973).

Uehara, Y. and K. Suyama: Visualization of the adventitial aspect of the vascular smooth muscle cells under the scanning electron microscope. *J. Electron Microsc.* **27**: 157—159 (1978).

Umetani, Y.: Postcapillary venule in rabbit tonsil and entry of lymphocytes into its endothelium: A scanning and transmission electron microscope study. *Arch. histol. jap.* **40**: 77—94 (1977).

Weiss, L.: The structure of bone marrow. Functional inter-relationships of vascular and hematopoietic compartments in experimental hemolytic anemia: an electron microscopic study. *J. Morphol.* **117**: 476—538 (1965).

Weiss, L.: Transmural cellular passage in vascular sinuses of rat bone marrow. *Blood* **36**: 189—208 (1970).

Wisse, E.: An electron microscopic study of the fenestrated endothelial lining of rat liver sinusoids. *J. Ultrastr. Res.* **31**: 125—150 (1970).

Wright, J. H.: The histogenesis of the blood platelets. *J. Morphol.* **21**: 263—278 (1910).

Yamada, E.: The fine structure of the megakaryocyte in the mouse spleen. *Acta anat.* **29**: 267—290 (1957).

Zamboni, L. and D. C. Pease: The vascular bed of red bone marrow. *J. Ultrastr. Res.* **5**: 65—85 (1961).

Zimmermann, K. W.: Der feinere Bau der Blutcapillaren. *Z. Ges. Anat., Atb. 1*, **68**: 29—109 (1923).

CHAPTER III

DIGESTIVE SYSTEM

Oral Cavity (Tongue)

Plate III-1   Filiform Papillae of the Tongue.   Rabbit.

The upper surface of the tongue is covered by papillae filiformes whose dimension and structure differ conspicuously among species. In carnivores they are strong spines directed towards the throat, while in the human they are more weakly formed. In the tongue of the rabbit, which is shown here, they are intermediate in development.

Besides their function as a rasp on the tongue, the filiform papillae serve as *sensory* organs for touch sensation on the lingual surface, as sensory nerves are richly supplied to the base of each papilla.

B shows a closer surface view of a filiform papilla. It is covered by cornified squamous epithelium. The epithelial cells show well developed *microplicae*, whose pattern is reticular on the upper or anterior side of the papilla, while it is parallel on the bottom or posterior side. The small round bodies are oral cocci.

A: × 1,200,   B: × 11,000

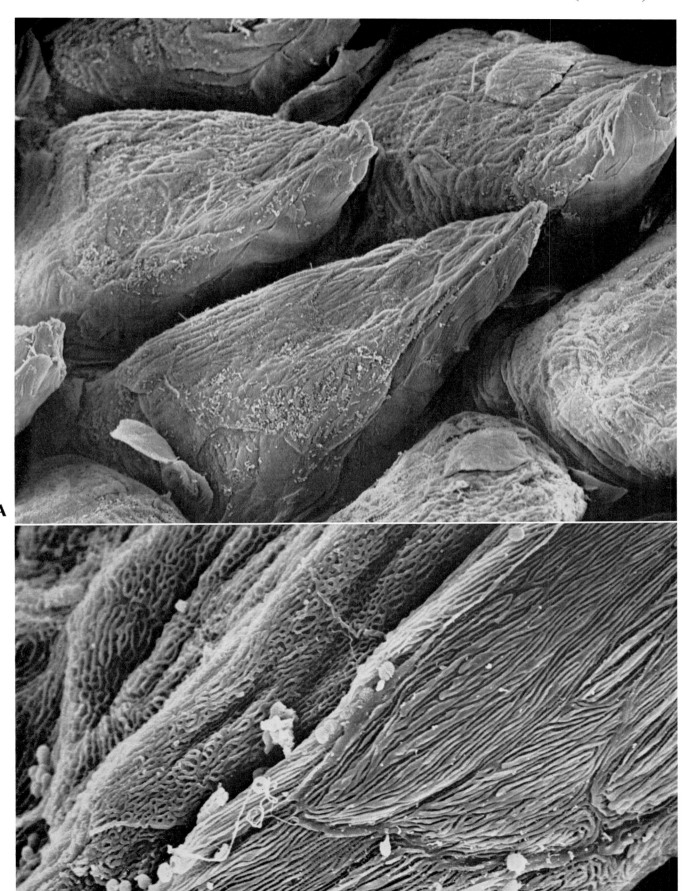

A

B

Plate III-2
A. Vallate Papilla of the Tongue. Rabbit.

The posterior boundary of the body of the tongue is provided with a row of papillae vallatae. As shown in this micrograph, the vallate papilla is a round disk surrounded by a deep groove, the *sulcus papillae*. The lateral epithelium of the disk, i.e. the internal wall of the groove, and its opposite epithelium, i.e. the external wall of the groove, are both populated with taste buds. A mass of salivary glands opens to the bottom and to the external wall of the groove to eliminate the taste substances from the groove.

   × 170

B. Taste Pores in Sulcus Papillae. Rabbit.

This micrograph shows the fairly regular distribution of *taste pores* on the external wall of the groove surrounding a vallate papilla. This view could fortunately be obtained as the papilla was removed in fracture.

   Each round pore is filled with the microvilli of taste bud cells. The SEM images of taste buds and their cellular components will be demonstrated in the chapter on sensory organs (Plates XI-14—16).

   × 3,000

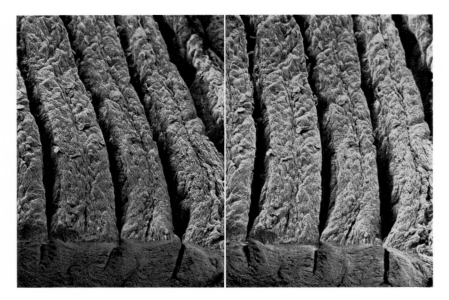

Fig. III-1 Foliate Papillae of the Tongue in Stereo. Rabbit.
The lateral edge of the tongue is equipped with parallel folds, papillae foliatae, which are especially well developed in the rabbit. The papillae extend crosswise to the longitudinal axis of the tongue. The lateral epithelium of each papilla is populated with taste buds.

   × 45

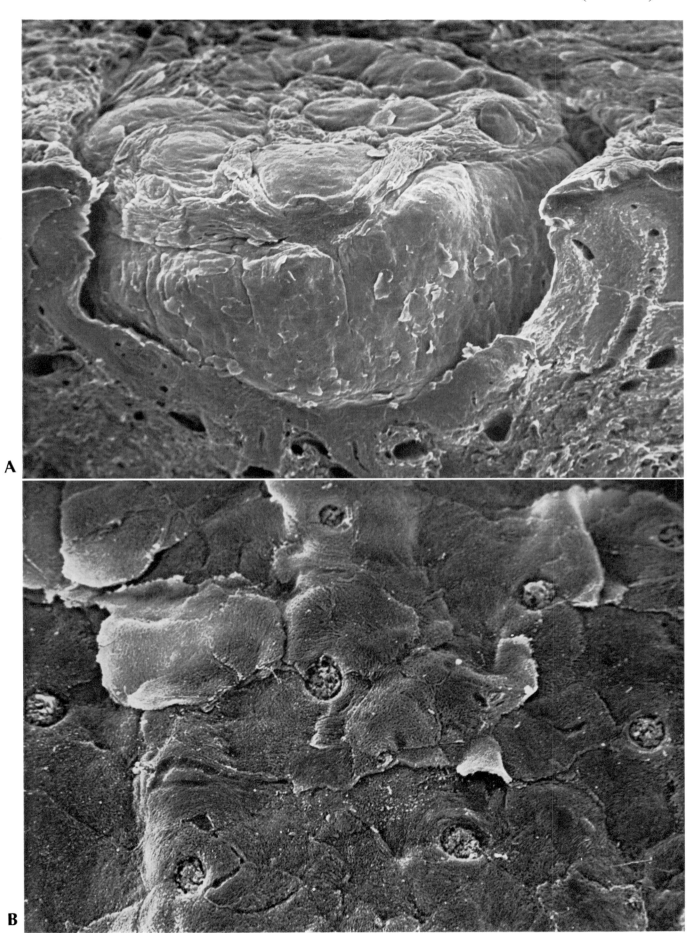

Esophagus

Plate III-3
A. Esophageal Epithelium.   Rabbit.

The squamous epithelium of the esophagus is fractured perpendicularly. The epithelial cells are generated in the *basal layer* where the cells and their nuclei are perpendicularly elongated. While the cells gradually shift towards the epithelial surface, they enlarge in cytoplasmic volume and flatten in a horizontal direction. Thus, 15—20 nucleated cells are piled up and, on the top, the nuclei disappear and strongly pressed sheets of cells form the superficial layer which is, in contrast to the corresponding layer in the epidermis of the skin, not cornified.

   The strings beneath the epithelium are collagen fibers in the lamina propria mucosae.

   X 740

B. Esophageal Surface.   Rabbit.

The superficial epithelial cells are covered by a reticular pattern of *microplicae* (Andrews, 1976). The cell boundaries usually are marked by a crest which is juxtaposed to the corresponding crest of the adjacent cell (arrows). The funnel-like pit in this micrograph seems to be the orifice of an esophageal gland and the columnar structure in the orifice is presumed to be the coagulated secretion of the gland.

   X 5,000

C. Esophageal Musculature.   Rabbit.

The tension and contractility of the esophageal wall is ascribed to the tunica muscularis which consists of skeletal muscle fibers in the upper part of the esophagus and of smooth muscle fibers in the lower part.

   This micrograph shows some skeletal (striated) muscle fibers, one cracked longitudinally, others crosswise. The sarcoplasm is largely lost during specimen preparation leaving myofibrils and parts of T-tubules. The cell membrane or sarcolemma is indicated by arrows, while N represents the nuclei of the muscle cells.

   X 640

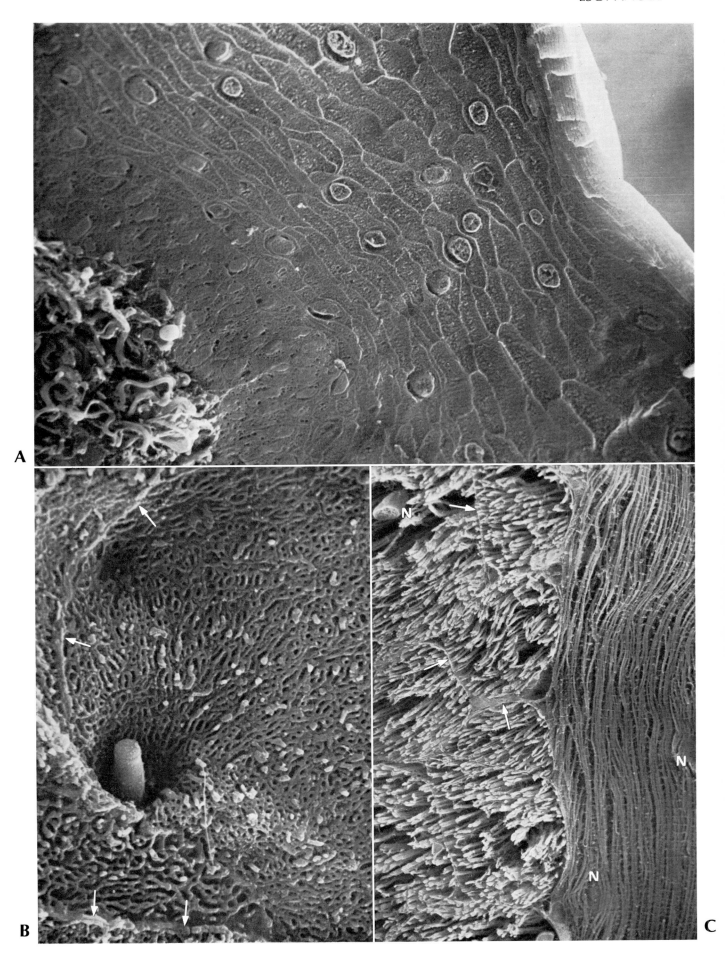

A

B

C

Stomach

The stomach of man and many mammals consists of the *gastric body* covered by fundic or oxyntic glands and of the *pyloric antrum* covered by pyloric glands. In some mammals the stomach shows a more complicated composition. The *rumen* of ruminants forms an independent sac and is covered by a specialized stratified squamous epithelium which has been observed by SEM by some authors. The oral half of the *murine* stomach is also lined by a stratified squamous epithelium as a continuation of the esophagus. Unfortunately, the human stomach is lacking in stratified epithelium.

The *oxyntic region* of the stomach shows dispersed *gastric pits.* The free surface and the pits are lined by surface mucous cells producing a vitally important mucus which protects the gastric surface from the ulcerogenic attack of hydrochloric acid. SEM images of the *mucus granules* released from the surface mucous cells in man have been published (Ogata and Murata, 1969). SEM analyses of these cells undergoing degranulation under different experimental conditions in animals seem to be a promising field.

The *fundic* or *oxyntic glands* are tubular in form; they are straight and only the bottom portion is distorted. A few of these tubes open to the bottom of each gastric pit. The three-dimensional construction of the gastric pit and associated fundic glands is impressively visualized by the SEM images of fracture surfaces produced by simply breaking the aldehyde fixed gastric mucosa (Hattori and Fujita, 1974; Takagi, 1974). Hattori and S. Fujita (1974) developed this "fractographic" study to apply to the analysis of the growth and multiplication of the fundic gland.

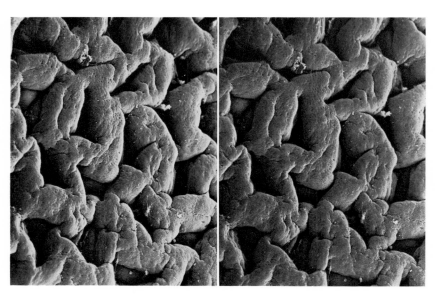

Fig. III-2 Gastric Surface in Stereo. Dog Oxyntic Area.
This low power view of the dog gastric fundus shows a honeycomb pattern on the mucosal surface with funnel-like grooves which correspond to gastric pits.
  X 120

The *fundic gland* consists mainly of *parietal cells* releasing hydrochloric acid and chief cells producing pepsinogen. The parietal cells posses *intracellular canalicules* covered by microvilli and a cytoplasmic system of finer tubulovesicles (for TEM studies, see Ito and Schofield, 1974). These structures, markedly changing their forms and numbers by the effect of gastrin, have been demonstrated by SEM (Osawa and Ogata, 1978) and will be interesting objects of study by the high resolution SEM.

The *pyloric antrum* also has pits but it is covered by cells differing from the surface mucous cells of the oxyntic area. *Pyloric glands* are mucous glands whose epithelium contains numerous endocrine cells producing gastrin and other gut hormones. SEM studies of these elements have not advanced as yet.

Plate III-4

A. Surface View of Gastric Mucosa.   Rat Oxyntic Area.

The oxyntic area of the stomach is covered by *surface mucous cells* whose apical faces are shown in this micrograph as polygonal and domed elements. Essentially no secretory signs are seen in this specimen obtained from a rat fasted for 24 hrs.

The deep, dark holes are the orifices of *gastric pits.*

X 740

B. Mucus Secretion after HCl Stimulus.   Rat Oxyntic Area.

It is known that hydrochloric acid experimentally introduced into the stomach, especially to the antropyloric region, conspicuously stimulates the mucus secretion from the mucous cells of the stomach. This suggests that gastric acid released into the lumen of the stomach promptly causes the release of mucus from the mucous cells. The significance of the *mucus barrier* in protecting the gastric mucosa against the ulcerogenic action of hydrochloric acid of the stomach is widely acknowledged.

This specimen was obtained from the rat stomach 5 min after administration of 3 ml of 0.1N HCl into the stomach. Each gastric pit extrudes a lacy belt of *mucus* which appears as such after the effect of fixation. The surface cells are partly perforated in the apical cytoplasm and partly expose their mucus granules (Fig. III-3, Page 110)

X 800

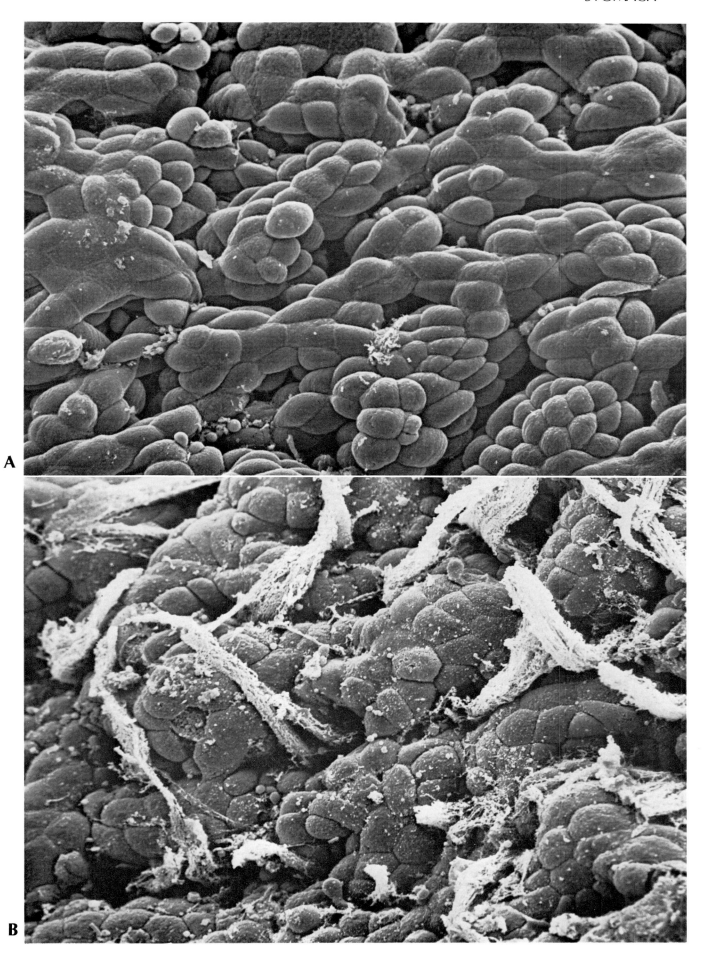

A

B

Plate III-5 Surface Mucous Cells. Rat Oxyntic Area.

The surface mucous cells contain mucus granules which appear characteristically flattened and tortuous in shape as observed in profiles by TEM. This SEM view of fractured cells shows that the granules are *concave discs* closely resembling erythrocytes in shape.

In the surface view of the cell apices, it is remarkable that the marginal zone of each cell is provided with short microvilli. As the surface mucous cells must respond very subtly to acid and other stimuli by releasing mucus, these microvilli may possibly represent the site of receptors of the cell.

N: nuclei and a cytoplasmic concavity which contained a nucleus.

X 14,000

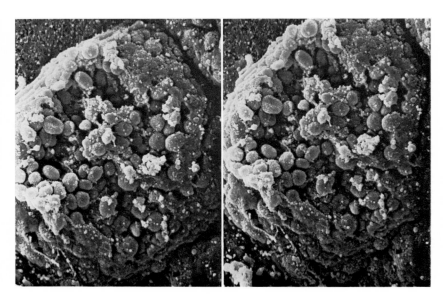

Fig. III-3 HCl Affected Surface Mucous Cell in Stereo. Rat

After HCl administration to the stomach, mucus secretion is conspicuously stimulated (Plate III-4B). In this state, the apical cytoplasm of the surface mucous cells becomes porous and it collapses in many cells, exposing more or less large masses of mucus granules as shown in this stereo-pair. The discoid shape of the granules is clearly recognized.

X 5,000

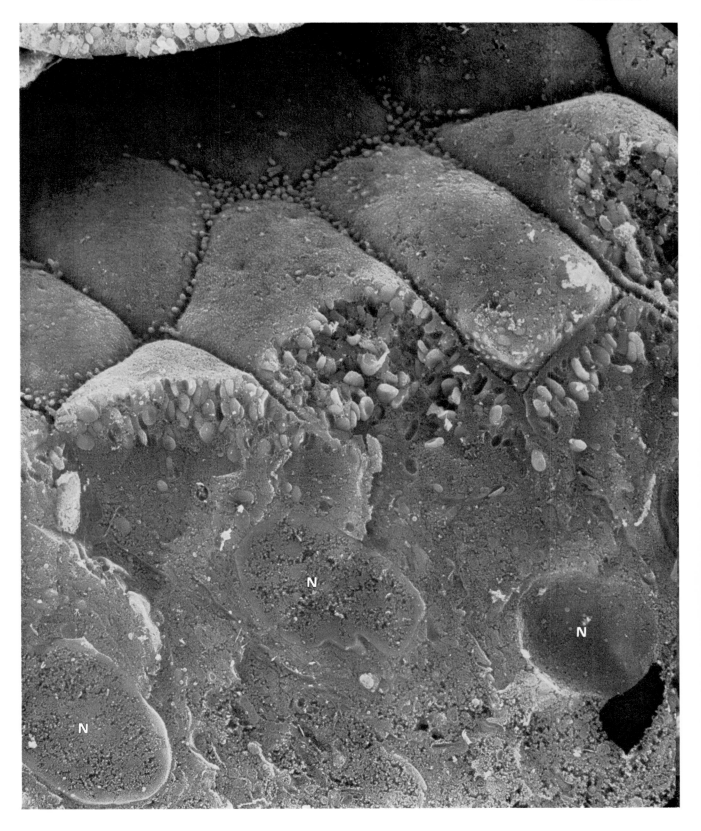

Plate III-6 Fundic Glands. Rat.

Fundic or oxyntic glands of the rat are horizontally cut by freeze fracture and also reveal their side aspect by mechanical tearing.

The *cross section* of each gland shows a small central lumen. The small tubular structures between the glands are blood capillaries, while the solid thinner bands are collagen fibers.

The *surface of the gland* shows some round swellings (∗) which correspond to *parietal cells*. A very delicate network of collagen fibrils (reticular fibers) surrounds each gland. Longitudinal thick collagen fibers are seen further outside.

X 700

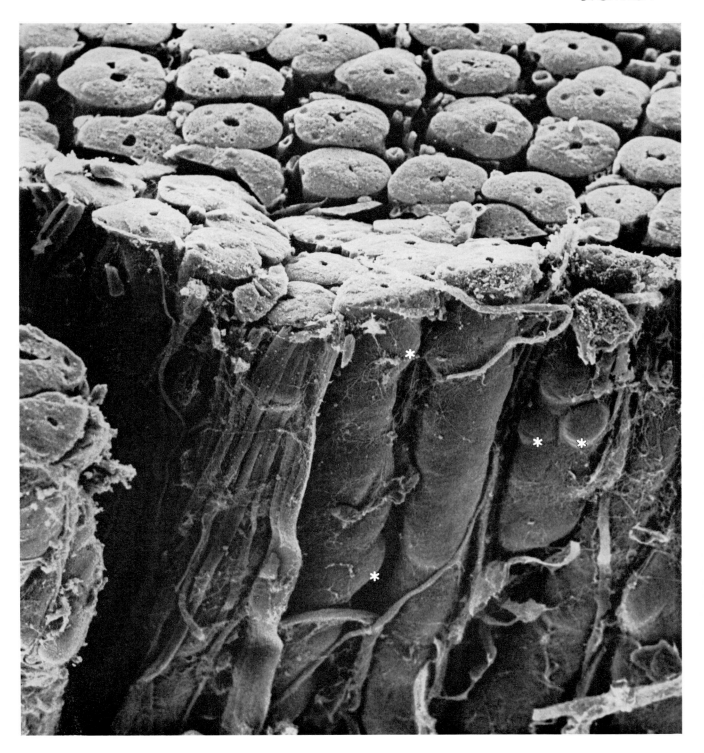

Plate III-7   Fundic Gland, Cross Cut.   Rat.

A fundic gland is horizontally fractured. The central lumen or the gland ductule is covered by small microvilli.

Large cells swelling outwards are *parietal cells* (P). Round mitochondria filling the cytoplasm are seen clearly as the cytoplasm in this specimen is rather poorly preserved.

Chief cells (C) are smaller and angular cells and show well developed lamellae of rough endoplasmic reticulum (ER). Their round secretory granules are not preserved in this specimen and appear as empty holes.

Blood capillaries and thick collagen bundles are seen in the interstitium. Delicate networks of collagen fibrils surround the fundic gland as well as the interstitial structure.

X 7,000

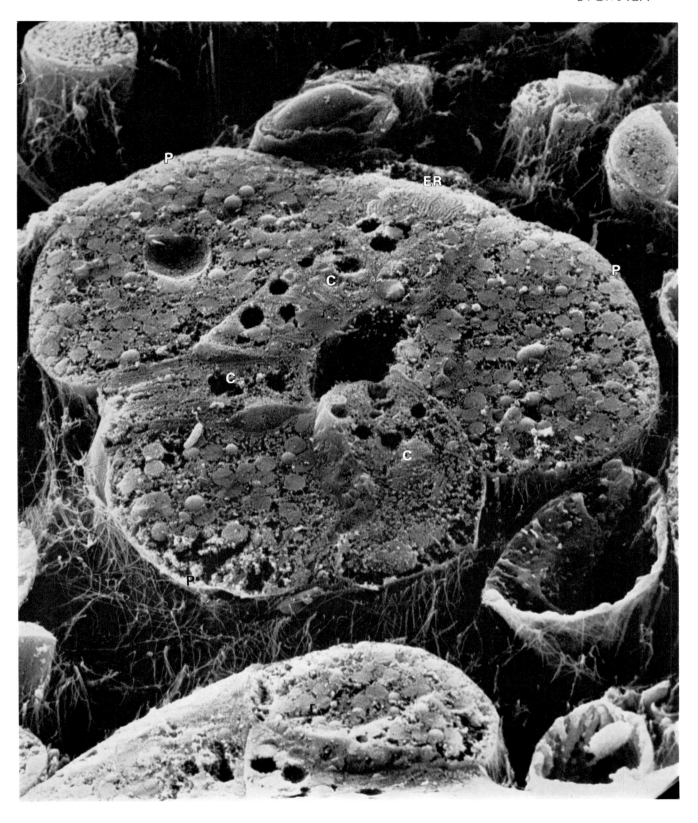

Plate III-8   Fundic Gland, Longitudinally Cut.   Rat.

A fundic gland is longitudinally fractured mainly revealing the cell boundaries and the cell interior in part. The *gland ductule* is opened revealing its surface covered by small microvilli.

The cells directly facing the ductule presumably are chief cells. One may see secretory material released from the cells (arrow).

Parietal cells (P) recede outwards and are supplied with a side branch of the ductule (*). One branch seems to receive the intracellular canalicules of two or three parietal cells.

X 3,700

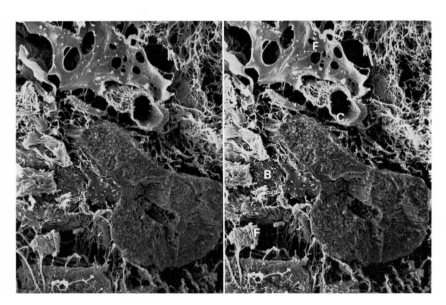

Fig. III-4 Lamina propria in Stereo.   Rat Stomach.
A fundic gland is cracked in the middle, while others reveal their basal surface covered by a basement membrane (B). The interstitium of the lamina propria mucosae represents a typical image of reticular tissue containing reticular fibers (delicate collagen bundles), fibroblasts (F) with attenuated extensions of cytoplasm and a blood capillary (C).

X 1,400

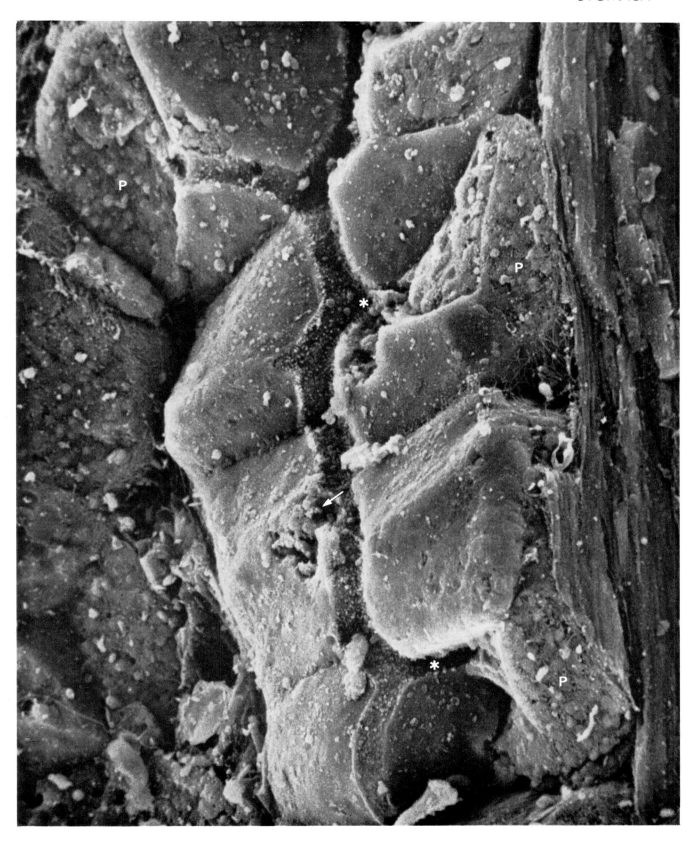

Plate III-9
A. Parietal Cell. Rat Stomach.

A subapical portion of parietal cell is shown. *Intracellular canalicules* (C) covered by large microvilli open to the apical space which is in the upper left of this micrograph. Numerous large round *mitochondria* are fractured in the cytoplasm revealing their dense cristae.

The tubulovesicular system (see page 107) is difficult to identify as the cytoplasmic matrix is slightly eroded in this specimen. This effect, however, serves as a clear demonstration of mitochondria.

   X 20,000

B. Orifice of Intracellular Canaliculus. Rat Stomach.

A parietal cell has been exposed by fracture. The antrum and orifice to the intracellular canalicular system are densely covered by irregular-shaped microvilli. Surrounding the orifice is seen a flat area for cell junction, dotted with small plateaus corresponding to *desmosomes* (D).

   X 32,000

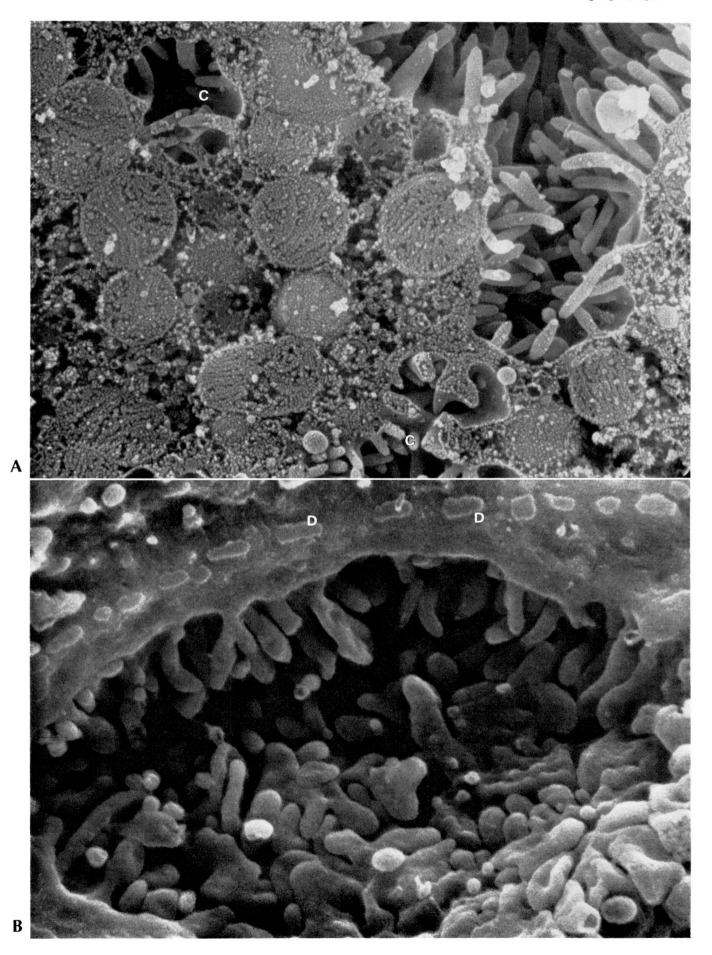

Plate III-10
A. Pyloric Mucous Cells.   Dog.

A pyloric pit is shown, which is the orifice of a pyloric gland. The *mucous cells* lining the surface are covered by short microvilli. The cell boundaries are markedly provided with a cytoplasmic ridge.

Essentially the same surface structure of pyloric epithelial cells has been reported in the monkey (Burke and Holland, 1976).

X 8,300

B. HCl Affected Pyloric Mucous Cells.   Dog.

The pyloric mucosal surface was examined 5 min after administration of 0.1N HCl into the pyloric region of the dog stomach. Many surface cells underwent degranulation with this treatment. The disk shape of the *mucus granules* is noticed. The cell surface shows numerous granular microprojections.

X 11,500

A

B

Intestine

The intestine has been observed by SEM by numerous researchers, especially because biopsy materials are easily applied. Yet the previous SEM studies of the gut have concentrated on the arrangement, shapes and surface structure of the *villi* and the orifices of *crypts* (Asquith *et al.*, 1970; Tayler and Anderson, 1972). Only a few papers are available concerning more advanced themata like the mechanism of cell loss from the villus top (Potten and Allen, 1977), developmental changes in the villi and their epithelium (Tsai and Overton, 1976; Lim and Low, 1977) and epithelial cell specialization in lymph follicles (Owen and Jones, 1974).

The fine structure and architecture of crypts and lamina propria mucosae (Plates III-15, 16) are unexplored regions for SEM and should be studied by advanced fracture and other techniques of specimen preparation.

Plate III-11 Surface View of Intestinal Villus. Human Duodenum.

The surface of a human duodenal villus is covered by columnar cells of absorptive function. Their apical surface, covered by *microvilli*, is either flat or convex and usually hexagonal in outline as is clear in this micrograph.

Dispersed among the absorptive cells are *goblet cells* with their characterstic tufts of microvilli.

X 2,400

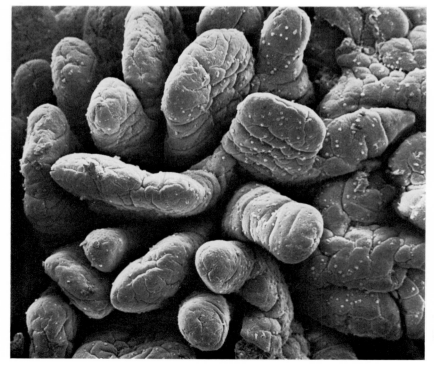

Fig. III-5 Low Power View of Human Duodenal Villi.
Biopsy.

Foliate and columnar villi are intermingled. The Clefts, which mainly run transversely on the villus surface and are familiar also in section preparations, have been ascribed to contraction of the villi by some authors (Demling *et al.*, 1969) and to rows of epithelial cells preformed shorter by other authors (Owen and Jones, 1974). Taylor and Anderson (1972) described these clefts as spiral in arrangement and discussed their relation to a possible spiral pathway of cell migration in the villus epithelium.

Note also the white dots on the villus surface which correspond to goblet cells.

X 75

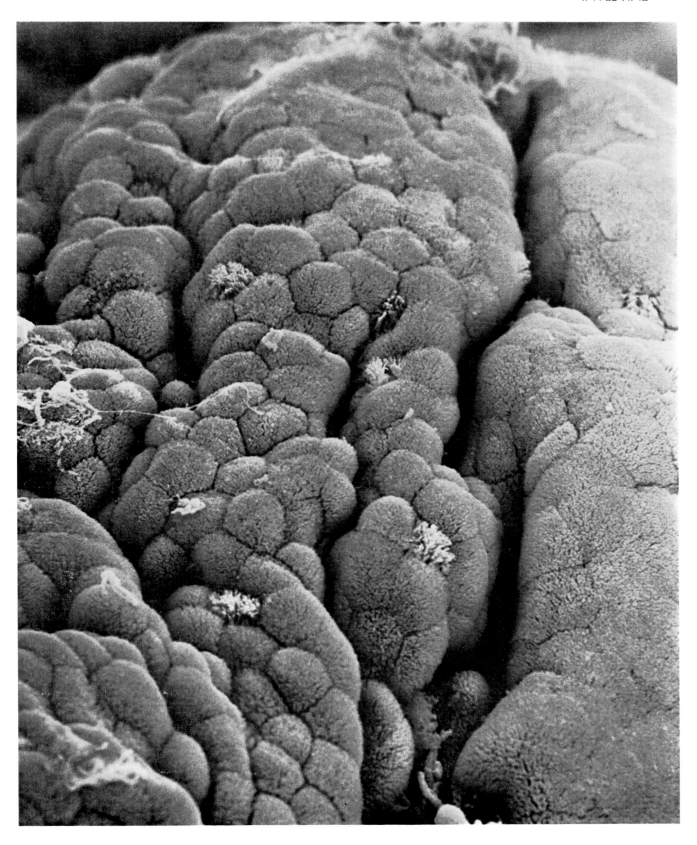

Plate III-12   Absorptive and Goblet Cells.   Rat Duodenum.

A. The epithelium covering a villus reveals both surface and fracture views. Fracture occured along the cell boundaries showing the prisms of absorptive cells. Three goblet cells (arrows) are intercalated among the absorptive cells.

B. Closer view of a goblet cell showing its apical dome with granular swellings and, in this case in contrast to the preceding plate, only a few microvilli.

C. Fracture surface of the epithelium showing the upper half of a goblet cell filled with granules of mucous secretions. They are dissolved and apparently being released through the narrowed apical end of the cell. The absorptive cells covered by microvilli show tortuous interdigitation (arrow).

A: X 3,000,   B: X 8,800,   C: X 8,500

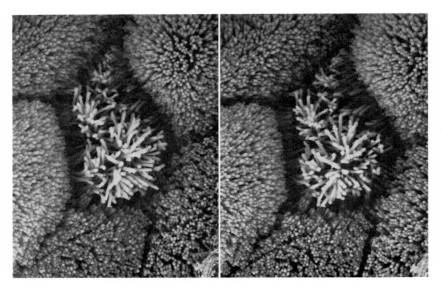

Fig. III-6 Goblet Cell Microvilli in Stereo.   Dog Duodenum.
A tuft of microvilli is surrounded by the microvilli of five absorptive cells.

Intestinal goblet cells conspicuously vary in surface structure according to different functional phases. In an apparently resting stage of the cell, the cell apex possesses a number of long microvilli as shown in this micrograph and in Plate III-11. These may likely serve as tentacles with which the cell subtly recognizes mechanical and chemical stimuli on the mucosal surface in order to respond to them by releasing its mucous secretions.

X 6,000

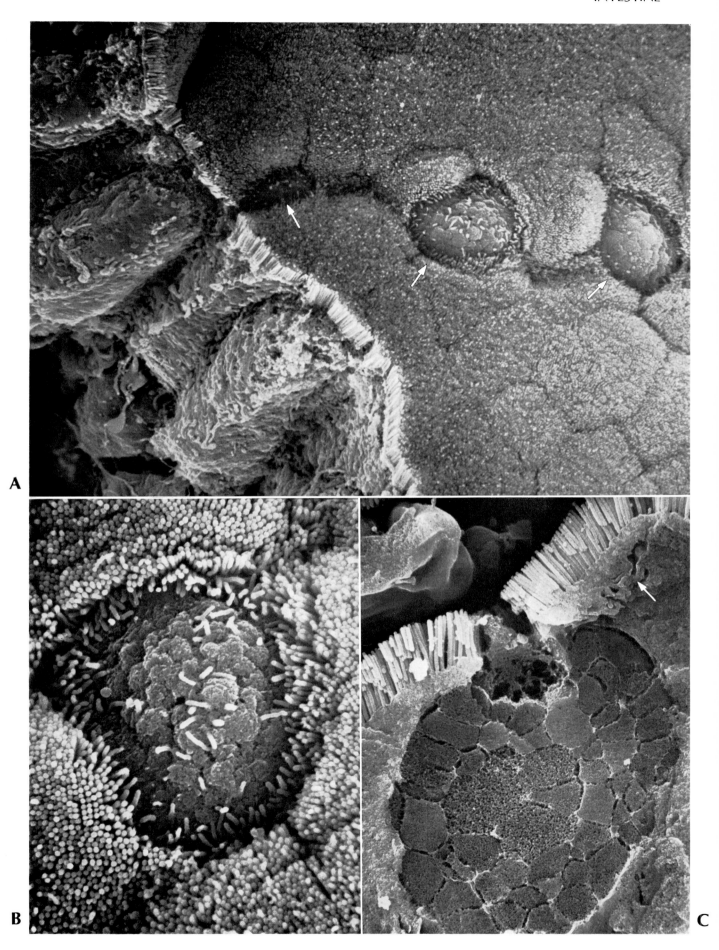

Plate III-13   Side View of Villus Epithelium.   Dog Duodenum.

Dog duodenal epithelium including a goblet cell (G) is perpendicularly fractured. Cell surfaces are partly revealed. The absorptive cells are complicatedly ruffled in the surface (*) to form laby-rinthine interdigitations. The cells, especially their basal portions, are markedly thinned, leaving rather ample *intercellular spaces*. Probable *lymphocytes* (L) are migrating through these spaces.

The *goblet cell* shows its typical shape, with a swollen mucus-containing portion, fractured, and a slender basal portion.

The *clefts of the villus* are formed by a short, middle cell (S) supporting the bottom of the furrow and, lateral to it, gradually longer cells with their apical portion sharply bent to cover the sides of the cleft.

C: blood capillaries in the lamina propria.

X 3,100

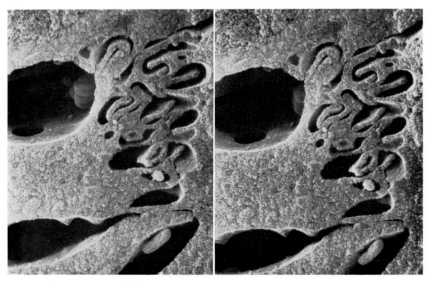

Fig. III-7 Epithelial Cell Interdigitation in Stereo. Dog Duodenum.
The absorptive cells at the base of the duodenal villus are fractured obliquely to the long axes of the cells. The cells are interdigitated with laminar microprocesses. Intercellular spaces, partly enlarged conspicuously, form a labyrinthine route for substance transport.

X 13,000

126

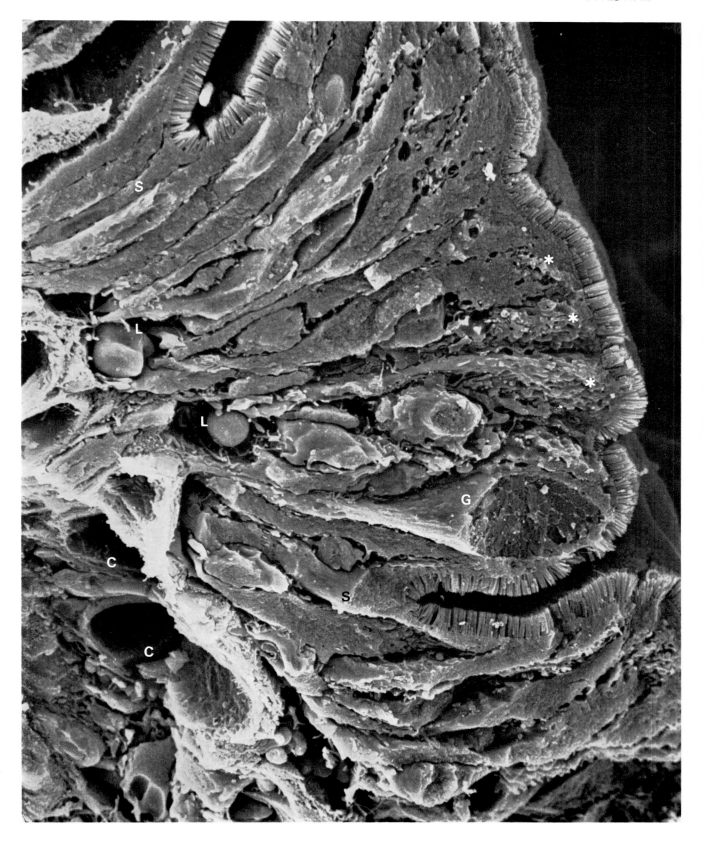

Plate III-14   Microvilli of Absorptive Epithelium. Dog (A) and Rat (B) Duodenum.

A. Microvilli and apical cytoplasm of absorptive cells are viewed from the side. The microvilli are fairly uniform in shape and size, measuring about 2 μm in length and 100 nm in thickness. Their surface is covered by fine granular material suggesting a remaining sugar coat.

B. Apical view of microvilli. Their uniformity looks more conspicuous than in the side view. The apparent sugar coat substance may be seen as spiny structures connecting adjacent microvilli.

The number of microvilli per cell differs considerably among different species and investigators: According to the TEM reports on the rat it may roughly range from 1000—3000 and according to a SEM study in humans, 2700—6500 (Reviewed in Taylor and Anderson, 1972). At any rate the increase in absorptive surface by the microvilli is enormous. Moreover, recent studies indicate that the microvilli move by the mechanism of actin and myosin (Mooseker and Tilney, 1975).

A: $\times$ 33,000,   B: $\times$ 33,000

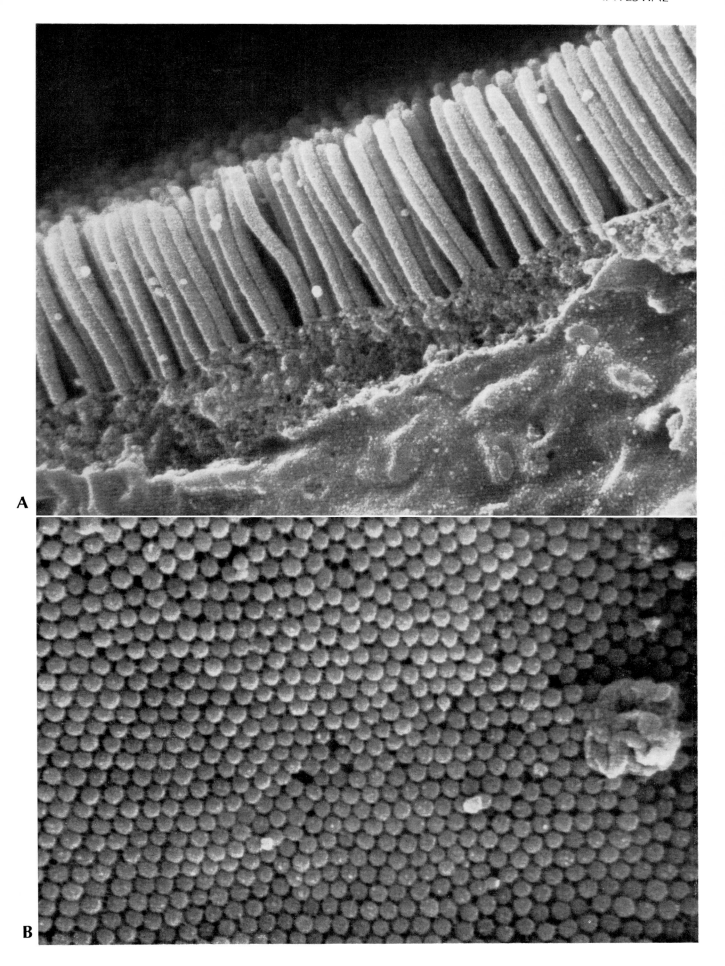

Plate III-15 Crypts and Lamina propria Mucosae. Dog Duodenum.

The intestinal crypts are also called Lieberkühn's glands and produce a large part of the intestinal fluid. The function and nature of the epithelial cells lining the crypts are still largely obscure.

In this micrograph the crypts are longitudinally fractured, revealing the narrow central lumen (arrows) and epithelial cells which include a few goblet cells (G). The intercellular spaces and interdigitations are seen. The microvilli of the cryptal cells are smaller than those of the villous ones.

Two blood capillaries and a probable venule (large space, bottom right) are opened. The lamina propria containing these vessels is a typical reticular tissue consisting of delicate reticular fibers, fibroblasts and wandering cells.

X 2,000

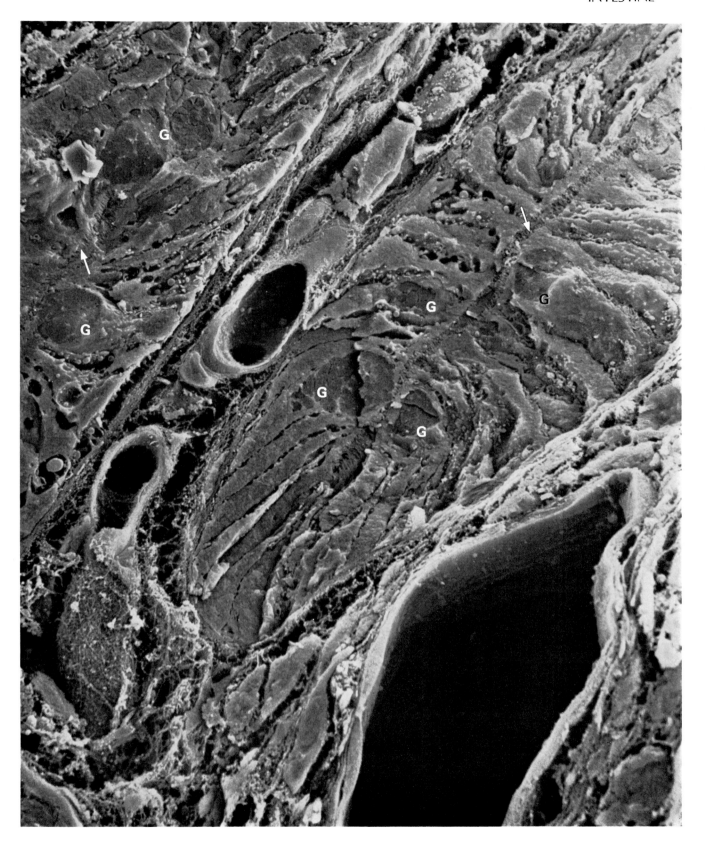

Plate III-16
A. Cryptal Epithelium.   Dog Duodenum.

Cells forming the crypt are fractured, mainly along the cell boundary. The cells are provided with apical microvilli and numerous plate-like projections which are interdigitated with those of the adjacent cells. Labyrinthine interdigitations are developed at the cell base.

    G: portions of goblet cells containing mucous granules.

    X 5,800

B. Flat View of Lamina propria.   Dog Duodenum.

The epithelial layer has been removed during the specimen preparation and the *basement membrane* and collagen fibrils of the lamina propria are exposed. The basement membrane looks like a ruffled sheet, revealing its apparently homogeneous material and very delicate filaments.

    X 12,000

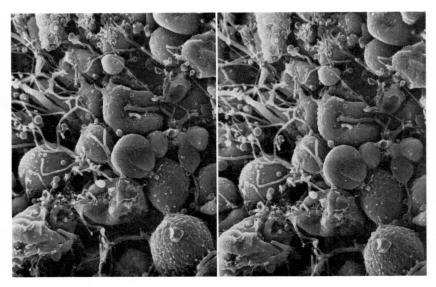

Fig. III-8 Lymphocyte Infiltration in Stereo. Dog Duodenum.
Lymphocyte infiltration occasionally is found in the lamina propria mucosae of the intestine. Reticular fibers loosely entangle the round elements which represent lymphocytes.

    X 3,000

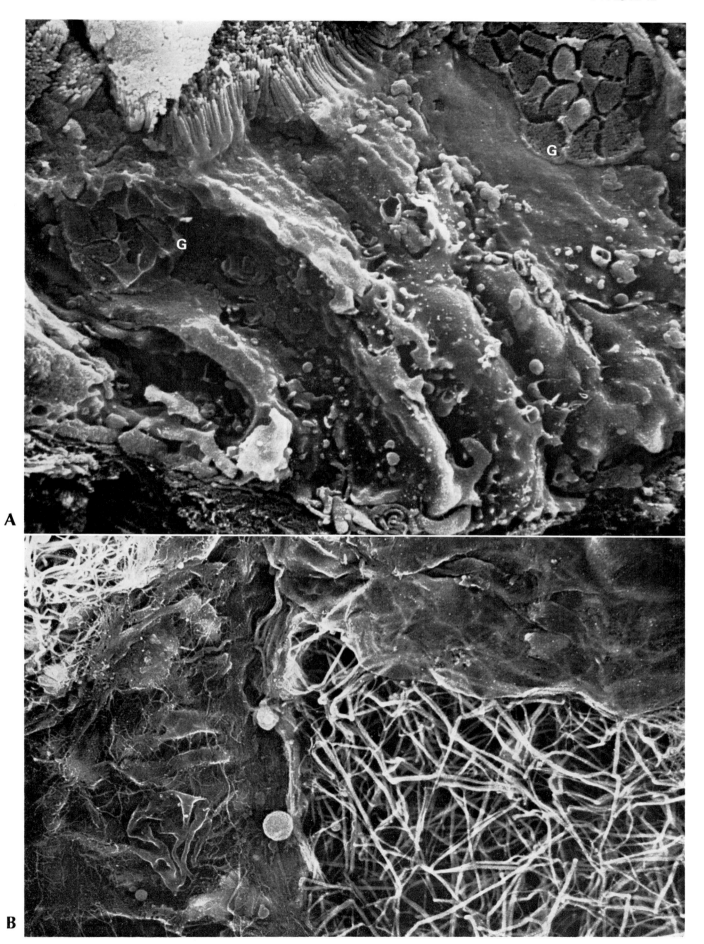

A

B

Colon

Plate III-17 Mucosal Surface of the Transverse Colon. Human.

The mucosal surface of the large intestine is composed of round *cryptal units,* each being delineated by a furrow and containing a crypt in the middle (Kavin *et al.,* 1970).

Each cryptal unit consists of concentrically arranged epithelial cells. Pits among these cells correspond to goblet cells whose mucus has been extruded. The boundaries and corners between the cryptal units show a rather loose arrangement of cells and these parts of the mucous surface likely correspond to the site where aged cells are gathered and abandoned.

X 600

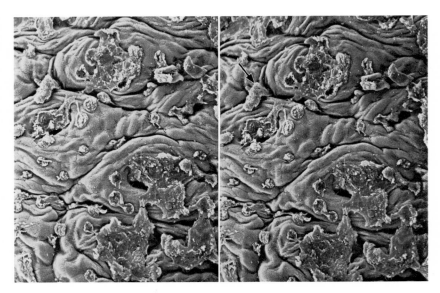

Fig. III-9 Cryptal Units in Stereo. Rat Colon.
The cryptal units are more clearly seen in the mucous surface of rat colon. In this specimen the mucous secretion from each goblet cell covers its neighboring cells. In the border of the cryptal units one may find a cell being separated from the epithelium (arrow).

X 400

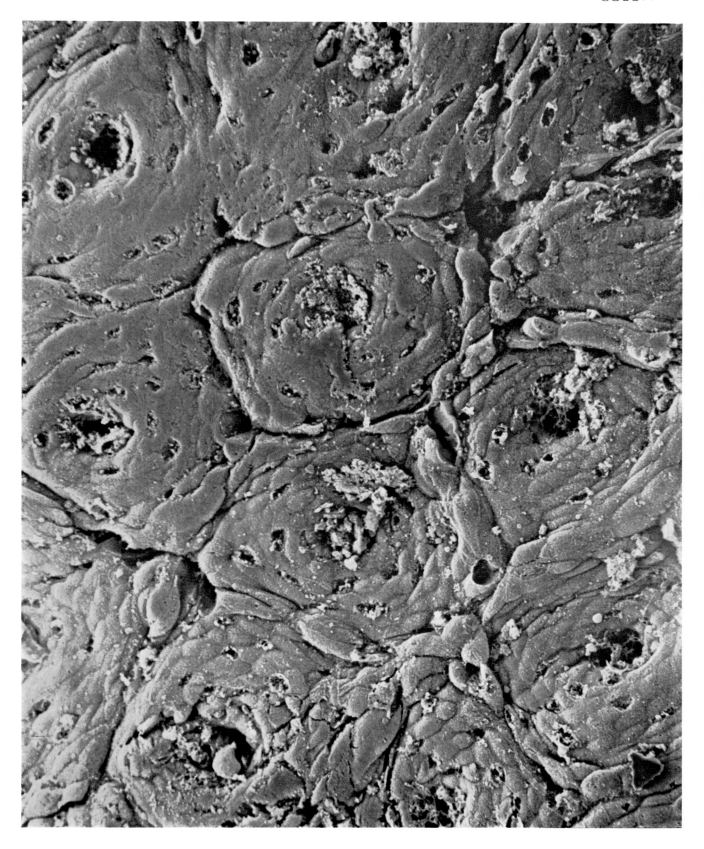

Pancreas (Exocrine)

The pancreas is a typical alveolar gland but its three-dimensional structure is by far more complex than usually depicted in histology textbooks. Moreover, the fine structure of different portions of the gland must change rapidly according to its functional dynamics. For the analysis of these problems SEM seems to be a useful weapon, though there are very few papers available demonstrating scientifically interesting SEM images of the pancreas (Motta *et al.*, 1977).

The following pages show the SEM images of the dog pancreas after 15 min stimulation by caerulein which is a mimic pancreozymin substance and a potent secretagogue of pancreatic enzymes.

Plate III-18   Acini and a Duct.   Dog.

In this specimen fracture occurred along the cell boundaries, and the *intercellular spaces* of the acini are visualized as dark channels. A *duct* (D) and a blood vessel (V) are in the middle. The outer surfaces of the acini and duct are covered by reticular fibers partly corresponding to the basement membrane.

    × 4,100

Plate III-19

A. Construction of Acinus. Dog Pancreas.

An acinus is cracked along the cell boundaries. The acinar cells are pyramidal in shape and form the glandular lumen bounded by their apices, which is by no means spacious; *intercellular secretory canaliculi* (arrows) extend towards the base of the acinus. These spaces are covered by microvilli, whereas the cell surface areas which faced the adjacent cell before fracture are undulated and partly provided with flat microprocesses.

   X 5,250

B. Intercellular Space of Acinus and Blood Capillary. Dog Pancreas.

An intercellular space of an acinus is opened by fracture. This space, though its relation to the intercellular canaliculi is obscure, is considerably enlarged apparently by the action of caerulein. The space extends to the acinar base (arrow), which, from the opposite side, is closely approached by a blood capillary.

It is widely accepted that pancreatic enzymes are released not only into the excretory duct of the organ but partly into the blood stream. (Occurrence of amylase in the blood is especially well known among physicians.) The blood levels of the enzymes are markedly increased after the pancreas is stimulated by pancreozymin or caerulein.

This SEM view probably visualizes how easily a leakage of pancreatic juice into the blood might take place especially when the pancreas is stimulated by its secretagogues. The site and function of the tight junction between the acinar cells, however, are not clear in this SEM image.

   * Fractured face of cytoplasm.

   X 4,400

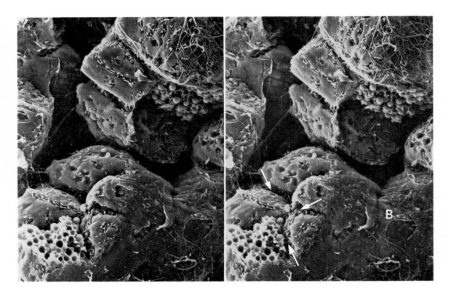

Fig. III-10 Pancreatic Acini in Stereo. Dog.
Cellular composition of the acini is visualized three-dimensionally. Some cells are cracked and round zymogen granules are exposed.

   The microvilli-covered grooves which, before fracture, formed the intercellular secretory canaliculi are indicated by arrows.

   The base of the acinus is uniquely undulated (Motta *et al.*, 1977) and provided with small microvilli. This surface is partly covered by a basement membrane (B) and, further from outside, by reticular fibers (upper right).

   X 1,650

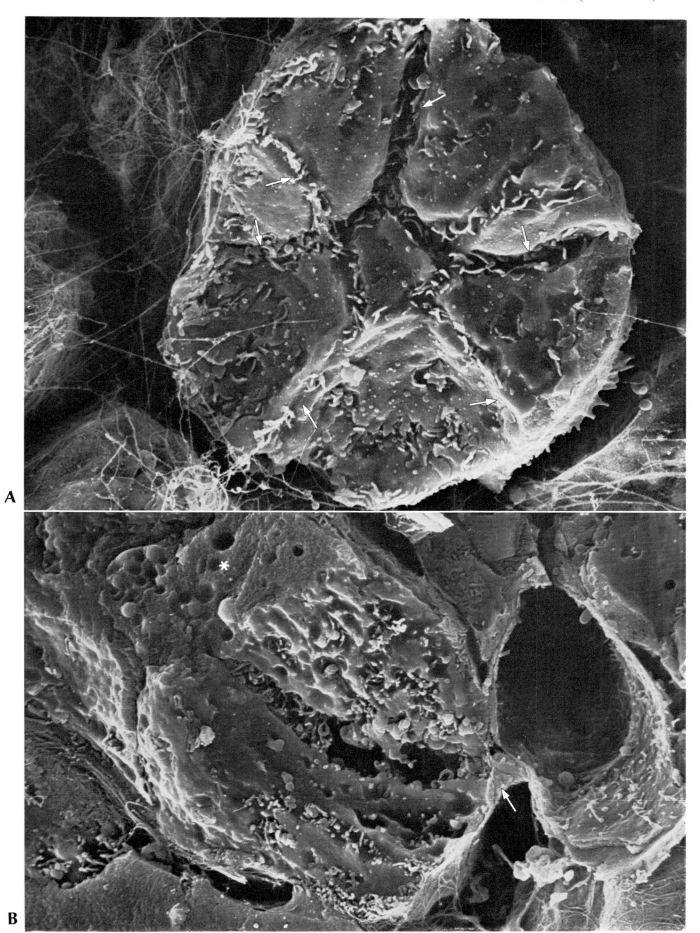

A

B

Plate III-20

A. Release of Zymogen Granules. Dog Pancreas.

In the pancreas stimulated by caerulein, numerous images of *granule release* (exocytosis) may be obtained under the SEM. The granule release may occur in the intercellular canaliculi but may be more easily identified as such on the smoother cell surface facing the intercellular clefts. This surface is believed to be sealed by a tight junction from the glandular lumen and intercellular canaliculi and, therefore, the granule release occurring here may possibly be "infrajunctional" and may account for the partial release of pancreatic enzymes into circulation.

Granules just being released are contained in a pit which corresponds to the granule sac opening to the cell surface. Some granules (thin arrows) are melting immediately after being released. Some other granules (thick arrows) are being released in tandem. Several granules are flattened and fused, being contained in a single exocytotic pit of the cell. The formation of the granules in tandem was first demonstrated by Ichikawa (1965) in his TEM observation of pancreozymin stimulated pancreas.

   * Fractured portions of the cell.

   X 16,000

B. Released Zymogen Granules. Dog Pancreas.

An *expanded intercellular space* between acinar cells contains numerous *granules* which are flattened into thin discs and piled up (thick arrows). On the right hand hollow granules are fused together like soap bubbles (thin arrows).

   X 16,500

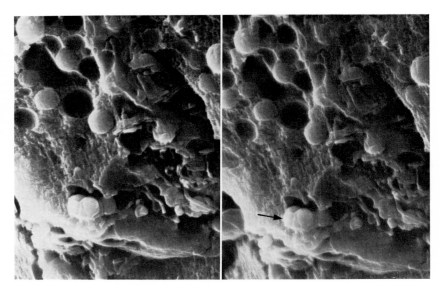

Fig. III-11 Zymogen Granules in Stereo. Dog. Pancreas.
An acinar cell is cracked and zymogen granules in the cell are exposed. Some granules are removed and their sacs remain as round grooves. A *tandem of granules* (arrow) is seen which is just being released into the intercellular canaliculus which appears as a dark tunnel in this micrograph.

   X 10,000

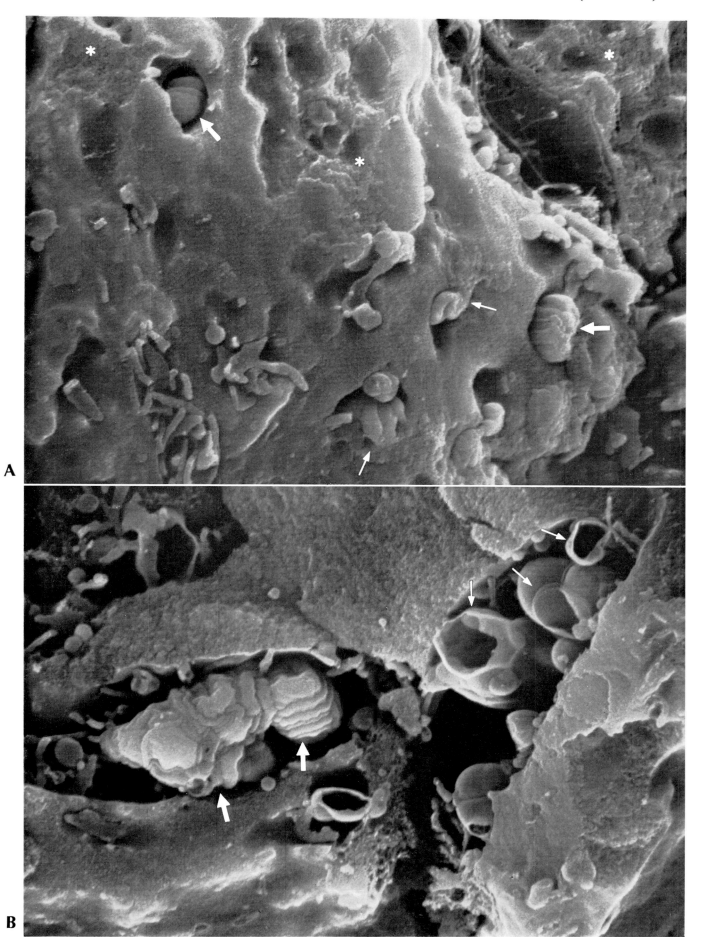

Plate III-21   Intercalated Portion.   Dog Pancreas.

The intercalated portion or isthmus is the ductule immediately proximal to the acini. Together with the centro-acinar cells which represent its direct extension into the acinus, the intercalated portion is generally believed to be the site of water and bicarbonate excretion, which is specifically stimulated by the gut hormone, secretin.

In this micrograph an intercalated portion is cracked crosswise precisely along the cell boundaries. Moreover, the *basement membrane* (B) has fortunately been peeled off and the basal face of the epithelium is exposed.

Reflecting the active function of water and bicarbonate release, both the luminal and basal surfaces of the epithelial cells are densely provided with microprojections. A central lumen of the ductule is rather inconspicuous and several intercellular canaliculi filled with microvilli are developed. The *basal microprojections* are mostly villous but partly laminar in shape. They are gathered along the base of cell boundaries, thus forming complicated interdigitations of the epithelial cells.

X 15,700

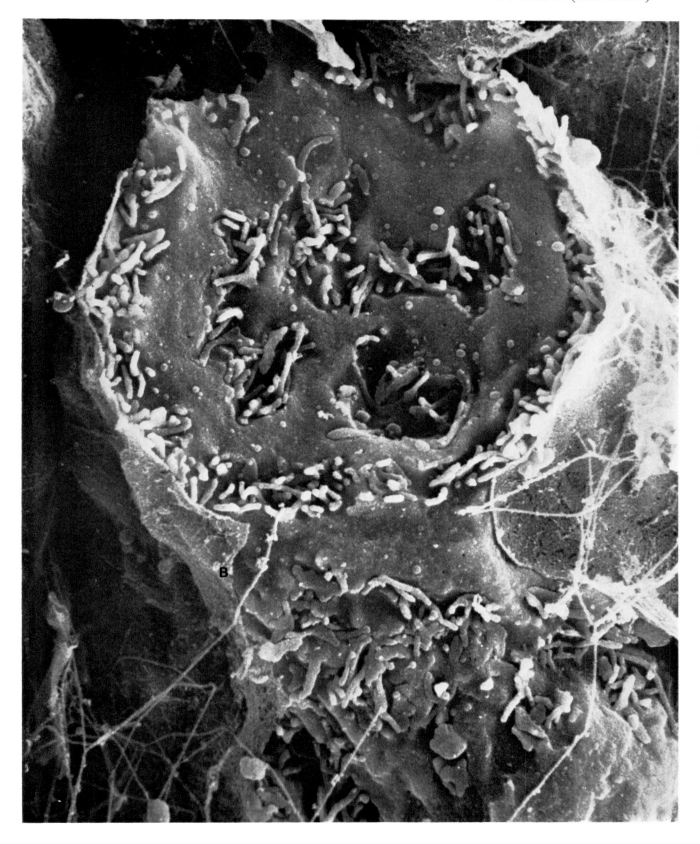

Liver

In the *liver lobule* the hepatocytes (liver cells) compose a network of plates which are caved and tunneled by sinusoids for blood flow. The cell plates are not two-cell-thick as early histologists thought but are one-cell-thick as Elias (1949) proposed by his stereological analysis of microscopic sections. The composition of this *"mularium simplex"* (Elias) and its three-dimensional extension are directly visible under the SEM.

The *hepatocyte* is roughly polyhedral in shape with different facets of different surface structures. The biliary face has a furrow for the bile capillary covered by microvilli and a smooth area on its both sides, corresponding to the junction against the adjacent cell. The sinusoidal face, lining the space of Disse, is covered by slender microvilli. SEM revealed that this space extends deep between the adjacent hepatocytes, thus forming an additional, rough facet to the biliary face of the hepatocyte (Motta and Porter, 1974; Motta *et al.,* 1978). It should be realized that the microvillous surface of a hepatocyte facing the space of Disse and its ample extensions represents a tremendously large area of substance transportation.

The attention of recent SEM investigators on the liver has been focused on the *endothelial fenestrations* in the sinusoids. Although Wisse (1970) proposed in his TEM studies that only small pores less than 0.1 μm are really existent and larger gaps are artifacts, more recent SEM observations after careful fixation and specimen preparation have confirmed the occurrence of both small pores and large fenestrations, including those exceeding 0.5 μm in diameter (Itoshima *et al.*, 1974; Grisham *et al.*, 1975, 1976; Motta, 1975; Muto, 1975; Motta *et al.*, 1978). SEM studies have evidenced that the endothelial gaps in liver sinusoids occur mainly within the cells or intracellularly, and only occasionally between the cells or intercellularly (Motta, 1975; Muto, 1975).

The morphology and nature of *Kupffer cells* have been much clarified by SEM. As macrophages in other organs, they are characteristically covered by cytoplasmic microprojections and there are found no gradations between these and endothelial cells. SEM studies thus have played an important role in denying the previous view that Kupffer cells might represent an activated stage of endothelial cells (Motta, 1975; Muto, 1975; Muto and Fujita, 1977).

Recent progress in the SEM studies of *vascular casts* has contributed much to the understanding of the microcirculation in the liver. Thus, the terminal ramifications of the interlobular arteries and different modes of their connection with the lobular sinusoids are being elucidated. Furthermore, the studies are concentrated on the re-examination and re-evaluation of the hitherto disputed peribiliary portal system (Murakami *et al.*, 1974; Motta *et al.*, 1978).

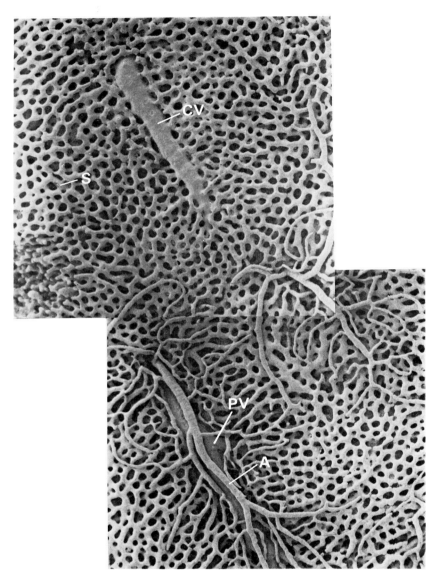

Fig. III-12 Vascular Cast of Liver Surface. Rhesus Monkey.
The fundamental design of the microcirculation in the liver may be visualized rather two-dimensionally in this vascular cast from the surface of the organ. The main vascular branches, i.e. the terminal branches of the hepatic arteries (A), and of the portal veins (PV) as well as the central veins (CV) appear just directly beneath the serous covering of the liver. Notice especially the delicate arborization of the arterial ends and their connections to the network of sinusoids (S).

   X 65

(Courtesy of Dr. T. Murakami, Department of Anatomy, Okayama University Medical School).

Plate III-22   Hepatic Lobule.   Human.

This micrograph shows a quarter of a hepatic lobule with the *portal tract* (interlobular tissue) at the bottom and a central vein in the upper right corner. The *hepatic plates* are twisted and anastomosed with each other and labyrinthic blood routes, sinusoidal capillaries, are formed between them. In this and the next micrographs, fracture occurred not along the cell boundaries but intracellularly, and therefore the bile capillaries are not clearly seen.

    CV: central vein, PV: interlobular vein, A: interlobular artery.

    × 400

(Courtesy of Dr. M. Muto, Department of Anatomy, Niigata University Medical School).

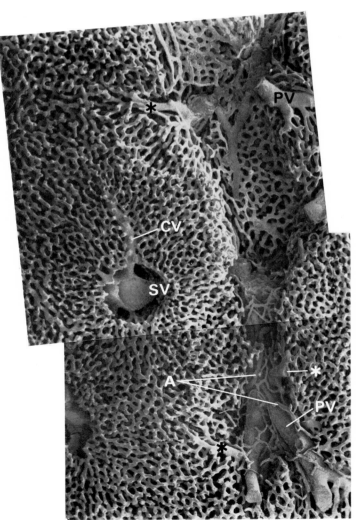

Fig. III-13   Vascular Cast of Liver Interior.
Rhesus Monkey.

The vascular cast of a liver was cut and partly microdissected to demonstrate the portal vein branches (PV, interlobular veins) and the hepatic artery branches (A, interlobular arteries). The former extend side branches (*) to the sinusoids of the lobule. The latter are ramified into delicate, twisted interlobular arterioles, corresponding to the view of the fractured interlobular tissue in Plate III-22. These arterioles are connected to the sinusoids in numerous places to give oxygen to the lobules.

    The sinusoidal blood of the lobule is collected by the central vein (CV) and conveyed to the sublobular vein (SV).

    × 65

(Courtesy of Dr. T. Murakamai, Department of Anatomy, Okayama University Medical School).

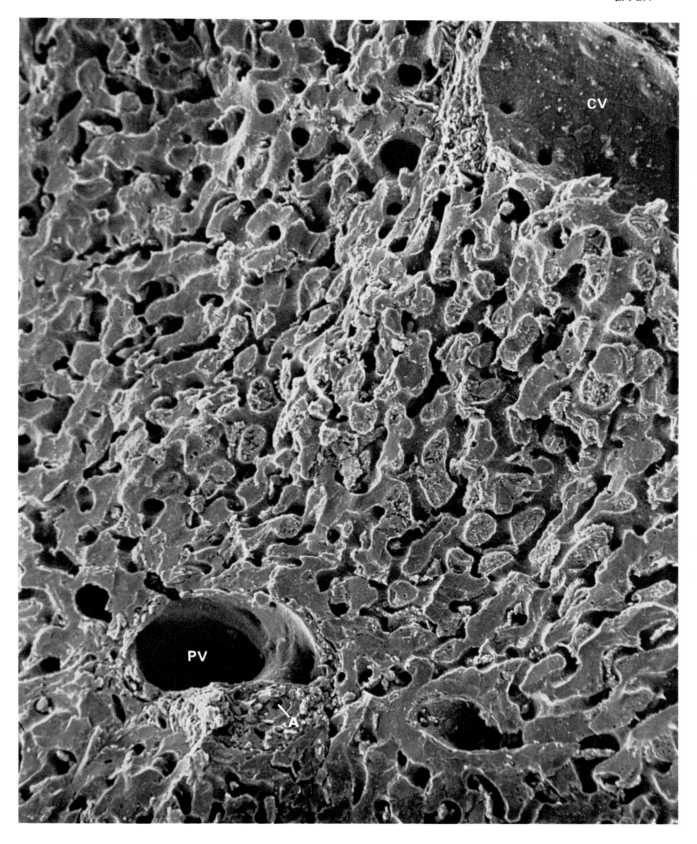

Plate III-23 Hepatocyte Plates and Sinusoids. Human.

The three-dimensional extension of the hepatocyte plates is shown in this micrograph. The spaces between them are sinusoids covered by a smooth and thin endothelium. The sinusoid surface is spotted with *Kupffer cells*, which are stellate and irregular in shape and densely covered by microprocesses. They protrude into or bridge across the sinusoid apparently to reach their prey coming with the blood more effectively. Some authors believe that Kupffer cells may control the sinusoidal blood flow by changing the degree of protrusion into the sinusoid.

 X 1,000

(Courtesy of Dr. M. Muto, Department of Anatomy, Niigata University Medical School).

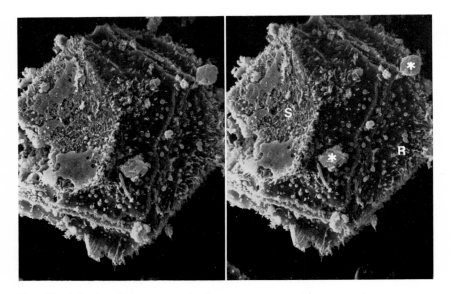

Fig. III-14 Isolated Hepatocyte in Stereo. Rat.

Critical point-dried specimens of the liver were crushed with forceps and fairly clean, isolated hepatocytes could be obtained. This micrograph shows a typical, polygonal hepatocyte.

 S is the *sinusoidal face* covered by microvilli. A part of the endothelial sheet remains attached. The microvillous surface extends around the cell edge towards the bile capillary. This surface (R), together with the corresponding surface of the opposite cell, forms an *intercellular recess* of the perisinusoidal space of Disse. Except for this rough area, the *biliary face* of the cell containing the bile capillary is essentially smooth. The nature of the globular bodies attached to this surface is unknown, though the possibility of their being unidentified secretions of the hepatocyte on this cell face may be worthwhile proposing.

 * Contaminants.

 X 2,000

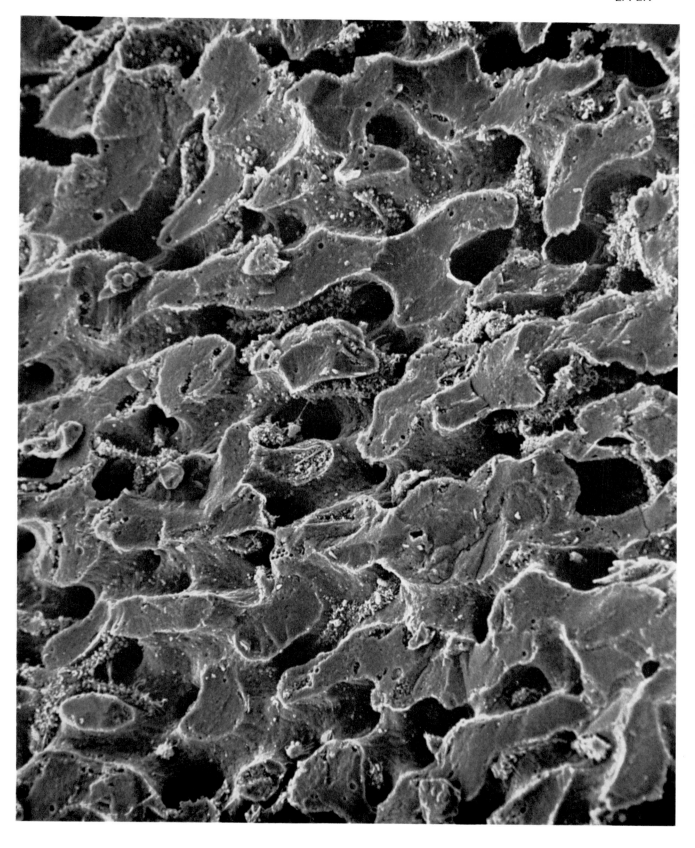

Plate III-24   Hepatocyte Plate and Bile Capillary.   Mouse.

This specimen is fractured along the boundaries of hepatocytes so that an overview of the courses of bile capillaries is given. Side branches of the capillary blindly end both between the cells and in the cell.

The *biliary face*, with the bile capillary in the middle, is flat, except for a few knob-like elevations and indentations corresponding to them (small arrows), indicating a type of cell interdigitation previously known from TEM studies.

Disse's space (D) is viewed in profile in this micrograph. As it continues into a space or recess between hepatocytes, it can be said that Disse's space extends deep into the hepatocyte plate. The *interhepatocytic recesses* (R), seen in this micrograph either in flat view or in side view, may reach closely to the bile capillary leaving a very narrow *zone of cell junction* (large arrow).

Thick and thin collagen fibers contained in the space of Disse are exposed in this micrograph. A cell regarded as either an endothelial or an Ito cell reveals its back (hepatocytic) side supported by collagen fibers (bottom right).

E: nuclear portion of endothelial cell.

X 3,500

(Courtesy of Dr. M. Muto, Department of Anatomy, Niigata University Medical School).

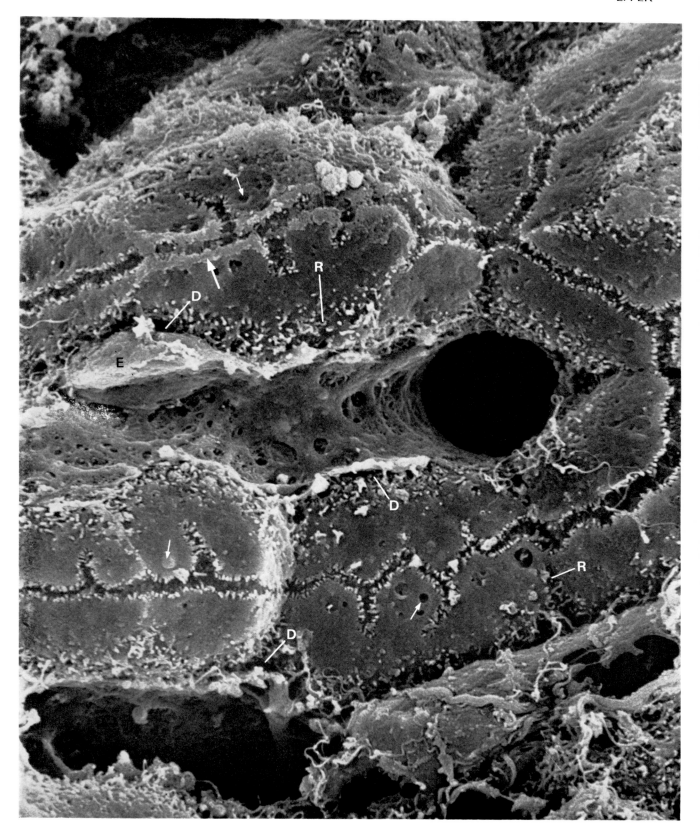

Plate III-25   Hepatocytes and Endothelium.   Human Liver.

The *biliary face* of hepatocytes is shown at higher magnification. This cell face is traversed by a bile capillary which is densely covered by uniform microvilli. On the right hand the bile capillary issues side branches.

A zone corresponding to the junctional complex of hepatocytes is not clear in this micrograph. (It is evident in Figure III-15.) The wide flat area of the biliary face shows broken holes. These are due to an interdigitation of hepatocytes by papillary processes and corresponding pits. One of the processes remains as indicated by the arrow.

The *space of Disse* (D) is filled with irregular microvilli covering the sinusoidal face of the hepatocytes. This rough face extends to form interhepatocytic recesses. A facet (R) covered by microvilli, though less densely than in the sinusoidal face, is thus identified on the biliary face.

The space of Disse is limited by the fenestrated endothelium of the sinusoid.

X 4,500

(Courtesy of Dr. M. Muto, Department of Anatomy, Niigata University Medical School).

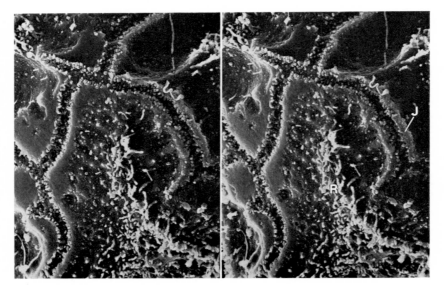

Fig. III-15 Biliary Face of Hepatocyte in Stereo.   Rat.
The biliary face of a hepatocyte is shown. Notice the smooth junctional zone (J) along the bile capillary, and the microvillous area (R) corresponding to an intercellular recess of Disse's space.

X 3,000

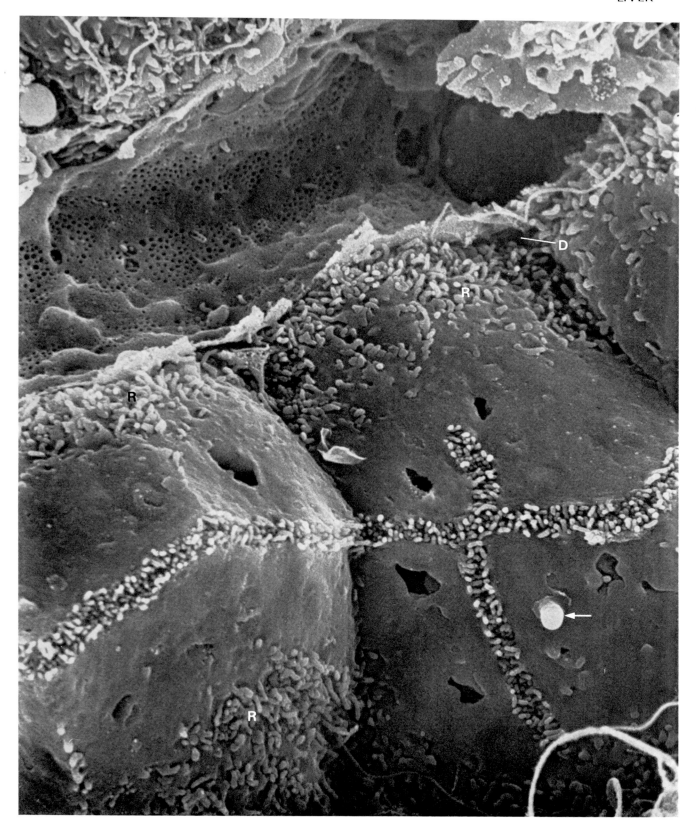

Plate III-26
A. Kupffer Cell. Human Liver.

A Kupffer cell extending three processes lies on the sinusoidal wall. Usually the endothelial sheet is lacking under this cell and thus the cell comprises a part of the lining of the sinusoid. Yet the Kupffer cell is quite different in structure from the endothelial cell. It is covered by microprojections of bulbous, villous and lamellar shapes. Long, tentacle-like ones may be called *filopodia* (F).

X 4,400

B. Kupffer and Endothelial Cells. Human Liver.

A part of a Kupffer cell surface covered by microprojections is shown. Some filopodia are attached to the endothelial surface.

The endothelial sheet is smooth and possesses large fenestrations and small, grouped pores. Underneath the endothelium is identified another sheet of a process-rich cell which is believed to be the fat-storing cell of Ito.

X 8,400

(Plat III-26: Courtesy of Dr. M. Muto, Department of Anatomy, Niigata University Medical School).

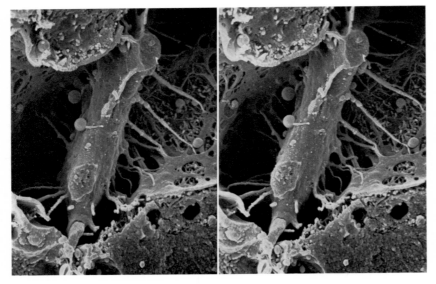

Fig. III-16 Kupffer Cell in Stereo. Rat.
A Kupffer cell extends its processes which overbridge the lumen of the sinusoid. Numerous filopodia anchor the cell to the endothelium which possesses oval fenestrations. Subendothelial collagen fibrils and Ito cell processes are seen under the endothelial layer.

X 3,300

(Courtesy of Dr. M. Muto, Department of Anatomy, Niigata University Medical School).

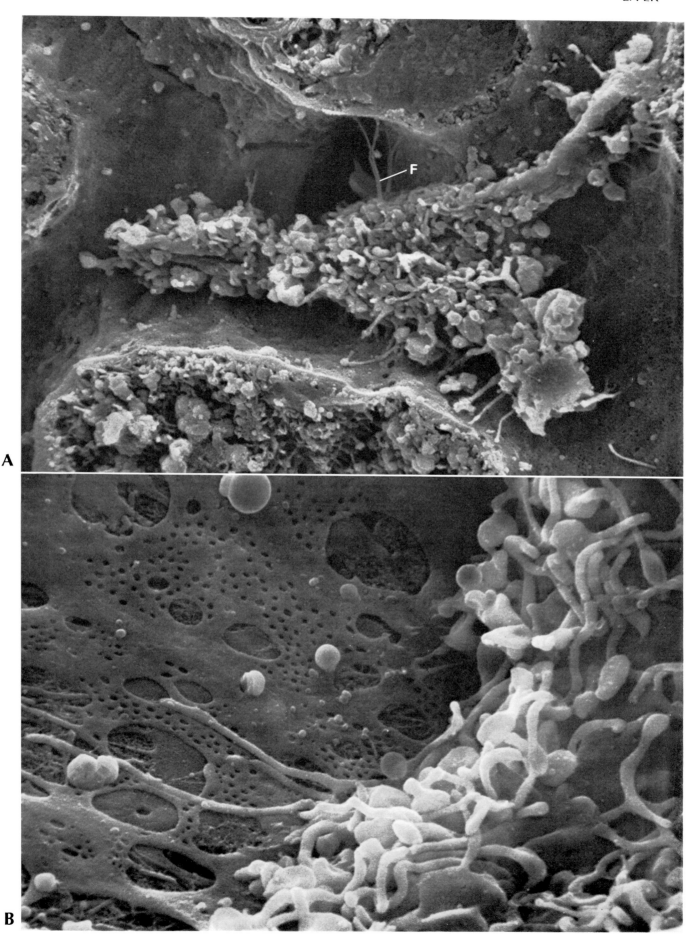

A

B

Plate III-27
A. Ito Cell Processes and Hepatocyte Microvilli.   Rat.

When the endothelial sheet (E) is peeled off in cracking, the sinusoidal face of a hepatocyte covered by slender and tortuous microvilli is exposed. Covering this face incompletely, however, another type of cell is found extending its flattened and complicatedly branching cytoplasm. This cell is an *Ito cell*, known also by the name of *fat-storing cell* as it characteristically contains fat droplets (Ito and Nemoto, 1952). The probable functions of this cell in collagen formation and in vitamin A storage and transport have attracted the attention of recent researchers (Wake, 1971; Ito, 1973; Kobayashi *et al.*, 1973).

X 16,500

B. Erythrocytes Phagocytosed by Kupffer Cell.   Rat.

The Kupffer cell is a *macrophage in the liver* specializing in phagocytosis of large bodies like aged or foreign erythrocytes. Endothelial cells of the sinusoid may incorporate small foreign particles but never such large bodies (Muto and Fujita, 1977).

Rat erythrocytes fixed for a short time in aldehyde were introduced into the portal vein. This micrograph derives from the liver specimen fixed 5 min thereafter and shows the erythrocytes captured by a Kupffer cell. They are entangled by its filopodia and partly swallowed by its cytoplasmic lips.

Erythrocytes with altered surface nature are detected by the filopodia of the Kupffer cell within a few minutes (Plate I-2B) and then attracted by them to the cell body. The cell consequently extends lip-like projections along the erythrocyte surface and they gradually cover the entire erythrocyte. This process of internalization of foreign bodies by Kupffer cell has been demonstrated by SEM both *in vitro* (Munthe-Kaas *et al.*, 1976) and *in vivo* (Muto and Fujita, 1977).

X 1,200

(Plate III-27: Courtesy of Dr. M. Muto, Department of Anatomy, Niigata University Medical School).

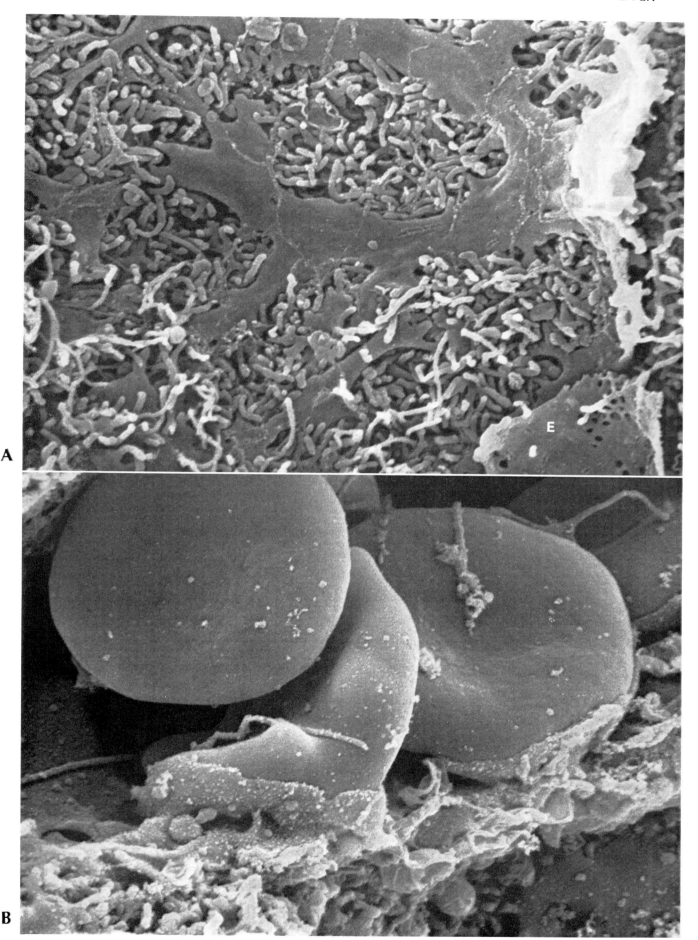

A

B

Gall Bladder

The gall bladder is generally known as the organ of bile concentration and secretion of mucous substances into the bile. Yet little is known concerning the precise functions and their control mechanisms. The gall bladder is phylogenetically quite an old organ, yet it may be nonexistent even in some familiar mammalian species such as the rat.

The mucous membrane of the gall bladder, rich in foldings, is covered by a single columnar epithelium which consists of mucus producing cells. Besides this type of cell, a few thinner rod-shaped cells and basal cells have been reported to occur in the human gall bladder.

Plate III-28   Gall Bladder Epithelium.   Rabbit.

Human gall bladder has been observed under the SEM by Laitio and Nevalainen (1972) and by Nielsen *et al.* (1975). In agreement with their findings, but under more complete elimination of mucous substances from the surface, this micrograph shows the domed apices of the mucus producing cells covered by low and uniform *microvilli*. It is known that mucous secretory granules are stored immediately beneath the domes.

   X 20,000

Plate III-29   Secretory Phases of Gall Bladder Epithelium.   Rabbit.

Careful examination of the gall bladder epithelium in untreated rabbits reveals structural changes in the cell apices which are believed to represent different *secretory* phases of the epithelial cells.

A demonstrates cells whose apical portion has lost microvilli and swells up to form *apocrine processes*. Some processes possess long, twisted microvilli.

B very likely shows the stage after a rupture of the apocrine processes containing mucous secretions. Some cells apparently keep the membranes of the cell process, while others show a large orifice through which mucous secretions in the cell body seem to be escaping.

A: X 15,000,   B: X 15,000

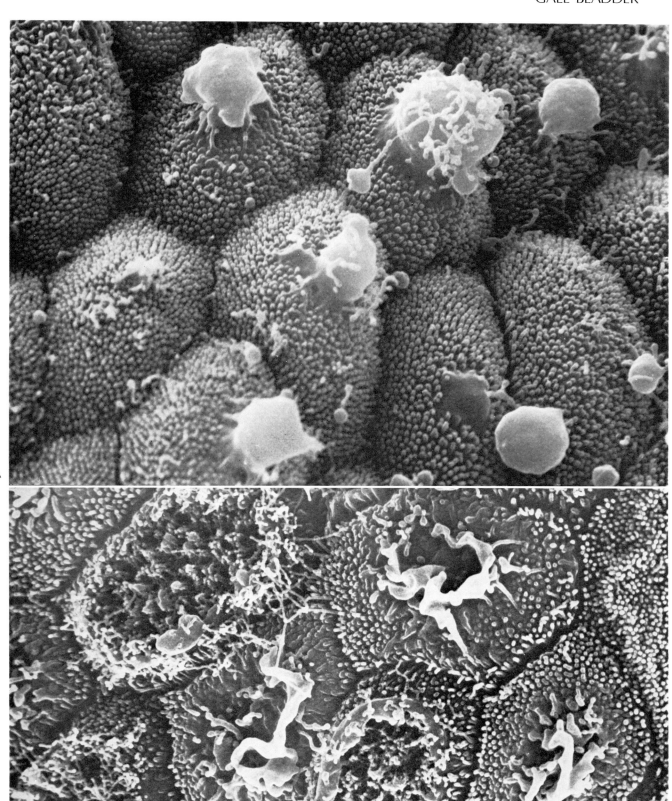

A

B

References

Andrews, P. M.: Microplicae: Characteristic ridge-like folds of the plasmalemma. *J. Cell Biol.* **68**: 420—429 (1976).

Asquith, P., A. G. Johnson and W. T. Cooke: Scanning electron microscopy of normal and celiac jejunal mucosa. *Amer. J. Dig. Dis.* **15**: 511—521 (1970).

Burke, J. A. and P. Holland: The epithelial surface of the monkey gastrointestinal tract. A scanning electron microscopic study. *Digestion* **14**: 68—76 (1976).

Demling, L., V. Becker and M. Classen: Examinations of the mucosa of the small intestine with the scanning electron microscope. *Digestion* **2**: 51—60 (1969).

Elias, H.: A re-examination of the structure of the mammalian liver. II. The hepatic lobule and its relation to the vascular and biliary systems. *Amer. J. Anat.* **85**: 379—465 (1949).

Grisham, J. W., W. Nopanitaya and J. Compagno: Scanning electron microscopy of the liver: A review of methods and results. In: (ed. by) H. Popper and F. Schaffner: Progress in Liver Diseases. Grune and Stratton, New York—London, 1975 (p. 1—23).

Grisham, J. W., W. Nopanitaya, J. Compagno and A. E. H. Nägel: Scanning electron microscopy of normal rat liver: the surface structure of its cells and tissue components. *Amer. J. Anat.* **144**: 295—322 (1976).

Hattori, T. and S. Fujita: Fractographic study on the growth and multiplication of the gastric gland of the hamster. The gland division cycle. *Cell Tiss. Res.* **153**: 145—149 (1974).

Ichikawa, A.: Fine structural changes in response to hormonal stimulation of the perfused canine pancreas. *J. Cell Biol.* **24**: 369—385 (1965).

Ito, S. and G. C. Schofield: Studies on the depletion and accumulation of microvilli and changes in the tubulovesicular compartment of mouse parietal cells in relation to gastric acid secretion. *J. Cell Biol.* **63**: 364—382 (1974).

Ito, T.: Recent advances in the study on the fine structure of the hepatic sinusoidal wall. A review. *Gunma Rep. Med. Sci.* **6**: 119—163 (1973).

Ito, T. and M. Nemoto: Über die Kupfferschen Sternzellen und die "Fettspeicherungszellen" ("fat storing cells") in der Blutkapillarenwand der menschlichen Leber. *Fol. anat. jap.* **24**: 243—258 (1952).

Itoshima, T., T. Kobayashi, Y. Shimada and T. Murakami: Fenestrated endothelium of the liver sinusoids of the guinea pig as revealed by scanning electron microscopy. *Arch. histol. jap.* **37**: 15—24 (1974).

Kavin, H., D. G. Hamilton, R. E. Greasley, J. D. Eckert and G. Zuidema: Scanning electron microscopy. A new method in the study of rectal mucosa. *Gastroenterol.* **59**: 426—432 (1970).

Kobayashi, K., Y. Takahashi and S. Shibasaki: Cytological studies of fat-storing cells in the liver of rats given large doses of vitamin A. *Nature, New Biol.* **243**: 186—188 (1973).

Laitio, M and T. Nevalainen: Scanning and transmission electron microscope observations on human gallbladder epithelium. *Z. Anat. Entwickl.-Gesch.* **136**: 319—325 (1972).

Lim, S.-S. and F. N. Low: Scanning electron microscopy of the developing alimentary canal in the chick. *Amer. J. Anat.* **150**: 149—174 (1977).

Mooseker, M. S. and L. G. Tilney: Organization of an actin filament-membrane complex. Filament polarity and membrane attachment in the microvilli of intestinal epithelial cells. *J. Cell Biol.* **67**: 725—743 (1975).

Motta, P.: A scanning electron microscopic study of the rat liver sinusoid. Endothelial and Kupffer cells. *Cell Tiss. Res.* **164**: 371—385 (1975).

Motta, P., P. M. Andrews, F. Caramia and S. Correr: Scanning electron microscopy of dissociated pancreatic acinar cell surfaces. *Cell Tiss. Res.* **176**: 493—504 (1977).

Motta, P., M. Muto and T. Fujita: The Liver. An Atlas of Scanning Electron Microscopy. Igaku-Shoin, Tokyo, 1978.

Motta, P. and K. R. Porter: Structure of rat liver sinusoids and associated tissue spaces as revealed by scanning electron microscopy. *Cell Tiss. Res.* **148**: 111—125 (1974).

Munthe-Kaas, A. C., G. Kaplan and R. Seljelid: On the mechanism of internalization of opsonized particles by rat Kupffer cells in vitro. *Exp. Cell Res.* **103**: 201—212 (1976).

Murakami, T., T. Itoshima and Y. Shimada: Peribiliary portal system in the monkey liver as evidenced by the injection replica scanning electron microscope method. *Arch. histol. jap.* **37**: 245—260 (1974).

Muto, M.: A scanning electron microscopic study on endothelial cells and Kupffer cells in rat liver sinusoids. *Arch. histol. jap.* **37**: 369—386 (1975).

Muto, M. and T. Fujita: Phagocytotic activities of the Kupffer cell: A scanning electron microscope study. In: (ed.by) E.Wisse and D. L. Knook: Kupffer Cells and Other Liver Sinusoidal Cells. Elsevier/North-Holland, Amsterdam, 1977 (p. 109—119).

Nielsen, O. V., M. L. Nielsen and K. E. F. Lauritzen: Examination of the mucosa of the normal human gall bladder by scanning electron microscopy. *Micron* **5**: 281—291 (1975).

Ogata, T. and F. Murata: Scanning electron microscopic study on the rat gastric mucosa. *Tohoku J. exp. Med.* **99**: 65—71 (1969).

Osawa, W. and T. Ogata: A scanning electron microscopy study on the fractured rat parietal cells in resting state and after stimulation with tetragastrin. *Arch. histol. jap.* **41**: 141—155 (1978).

Owen, R. L. and A. L. Jones: Epithelial cell specialization within human Peyer's patches: An ultrastructural study of intestinal lymphoid follicles. *Gastroenterol.* **66**: 189—203 (1974).

Potten, C. S. and T. D. Allen: Ultrastructure of cell loss in intestinal mucosa. *J. Ultrastr. Res.* **60**: 272—277 (1977).

Takagi, T.: Scanning electron microscopical studies of human gastric mucosa; fetal, normal and various pathological conditions. *J. clin. Electron Microsc.* **7**: 83—101 (1974).

Taylor, A. B. and J. H. Anderson: Scanning electron microscope observations of mammalian intestinal villi, intervillus floor and crypt tubules. *Micron* **3**: 430—453 (1972).

Tsai, L.-J. and J. Overton: The relation between villus formation and the pattern of extracellular fibers as seen by scanning microscopy. *Devel. Biol.* **52**: 61—73 (1976).

Wake, K.: "Sternzellen" in the liver: perisinusoidal cells with special reference to storage of vitamin A. *Amer. J. Anat.* **132**: 429—462 (1971).

Wisse, E.: An electron microscopic study of the fenestrated endothelial lining of rat liver sinusoids. *J. Ultrastr. Res.* **31**: 125—150 (1970).

CHAPTER **IV**

RESPIRATORY SYSTEM

The air tract with its extensive free surface provides suitable materials for SEM studies, yet it is considerably difficult to obtain clean specimens free of mucous substances covering the epithelial surface.

The vestibule of the *nasal cavity* close to the naris is lined by a haired, stratified epithelium. Thicker cells covered by short microvilli appear more posteriorly. Further posteriorly, prismatic cells with longer and denser microvilli predominate, while ciliated cells appear here and there. The posterior, major part of the nasal cavity is lined by a ciliated epithelium, but here also more or less numerous non-ciliated, microvillous cells are intermingled. This mixed (ciliated and non-ciliated) epithelium extends into the pharynx and into the accessory nasal cavities.

The *olfactory region* covering a distinct, superior portion of the nasal cavity is characterized by the occurrence of olfactory cells and will be dealt with in the chapter on sensory organs (Chapter IX).

The *larynx* is lined by a squamous epithelium in mechanically responsible portions such as the epiglottis and vocal cords, whereas other portions of the larynx are covered by non-ciliated cells partly intermingled with ciliated cells and a variable number of goblet cells. The non-ciliated cells are prismatic cells covered by short microvilli and called transitional cells by otolaryngologists (Plate IV-4).

The *trachea* is covered by typical ciliated cells with long cilia beating upwards. Numerous goblet cells may be intermingled. A few cells with a small apical area equipped with a tuft of microvilli were described in the TEM study by Rhodin and Dalhman (1956) under the name of *brush cells* and were later observed by SEM (Andrews, 1974; Smolich *et al.*, 1977). Although the function of the brush cell has been much disputed (see Smolich *et al.*, 1977), the hypothesis that they represent a chemoreceptor of the trachea (Luciano *et al.*, 1968) seems most worthy of attention. If this hypothesis is tenable, the brush cells might represent a type of basal-granulated or chemoreceptor-like cell distributed in the tracheal and bronchial epithelium (Lauweryns and Cokelaere, 1973; Cutz *et al.*, 1975; Capella *et al.*, 1978).

SEM observations of the extrapulmonary airway have been reported by numerous authors (Greenwood and Holland, 1972; Andrews, 1974; Breipohl *et al.*, 1977). SEM studies on the development (Smolich *et al.*, 1977) and pathological changes (Ohyama *et al.*, 1977a,b) of the laryngeal and tracheal epithelia are available.

The *intrapulmonary airway*, represented by the bronchial tree and its twigs or bronchioles, is lined by a ciliated epithelium containing numerous goblet cells. As mentioned above, previous light microscopic and TEM studies revealed the occurrence of grouped or solitary *basal-granulated cells* open to the mucosal surface of the intrapulmonary bronchi and bronchioles (Tateishi, 1973; Cutz *et al.*, 1974; Capella *et al.*, 1978; Taira and Shibasaki, 1978). These recepto-secretory type cells probably function as detectors of substances, including the oxygen content, in the inspired air (Lauweryns and Cokelaere, 1973), and their distribution and apical structures deserve careful scrutiny by SEM.

The connection and luminal view of the *terminal* and *respiratory bronchioles, alveolar ducts* and *alveoli,* including the *Clara cell* conspicuously protruding with its non-ciliated apex in the terminal bronchiole and the occurrence of *alveolar pores* between juxtaposed alveoli have been clearly demonstrated by the SEM (Okada, 1969; Nowell and Tyler, 1971; Tyler *et al.,* 1971; Greenwood and Holland, 1972; Kuhn and Finke, 1972; Castleman *et al.,* 1975). These studies have further revealed the surface structures of the *squamous* (Type I) and *great* (Type II) *alveolar cells* as well as the *alveolar macrophage.* The *"alveolar brush cell"* with a tuft of microvilli projecting into the alveolus has been demonstrated by SEM (Hijiya, 1978) as well as by TEM, and has been presumed, as the tracheal cell of the same name (*vide supra*), to have a chemoreceptor function.

Only a few but valuable SEM observations are available concerning the development (Krause and Leeson, 1973; Ishii, 1977), comparative histology (Nowell *et al.,* 1970) and pathology (Ogata, 1975; Kuhn and Tavassoli, 1976; Mellick *et al.,* 1977; Hijiya, 1978) of the lung.

Nasal Cavity

Plate IV-1   Nasal Cavity.   Rat.

A. The mucosal surface close to the olfactory area is shown. The mucosal epithelium represents a mixture of ciliated and non-ciliated cells. Some ciliated cells have long curved cilia which may number 200–300 per cell, but others possess less numerous, short straight cilia and appear to be immature cells. The apical surface of the non-ciliated cells is slightly domed and they are densely covered by short microvilli.

Besides these cells, a few cactus-like cells (arrows) are seen. They project a columnar apical process into the nasal cavity which radiates slender sticks of microvilli. It is possible that these cells represent a variation of non-ciliated cells, but an attractive hypothesis is that they are involved in a specific function like chemoreception in addition to the olfactory cells.

B shows a cactus-like cell and immature ciliated cells in closer view. A small cell apex labeled an asterisk (*) may possibly belong to the cactus-like cell line.

A: X 7,800,   B: X 16,000

(Specimen provided by Prof. M. Ohyama, Department of Otolaryngology, Kagoshima University School of Medicine).

A

B

Larynx and Trachea

Plate IV-2   Laryngeal Ventriculus.   Dog.

A. The larynx, as in this specimen, is partly lined by squamous epithelium. The epithelial cells show unique and complicated surface microprojections, which mainly belong to the category of microplicae (P) but partly deserve the name of microvilli (V).

B shows a closer view of the cell surface. At numerous places a microplica projects a microvillus whose thickness is identical with the width of the plica. Independent microvilli are also numerous among the microplicae. These figures strengthen the hypothesis (page 20, 26) that microplicae and microvilli may be changeable into each other.

   A: X 6,200,   B: X 20,000

(Specimens provided by Prof. M. Ohyama, Department of Otolaryngology, Kagoshima University School of Medicine).

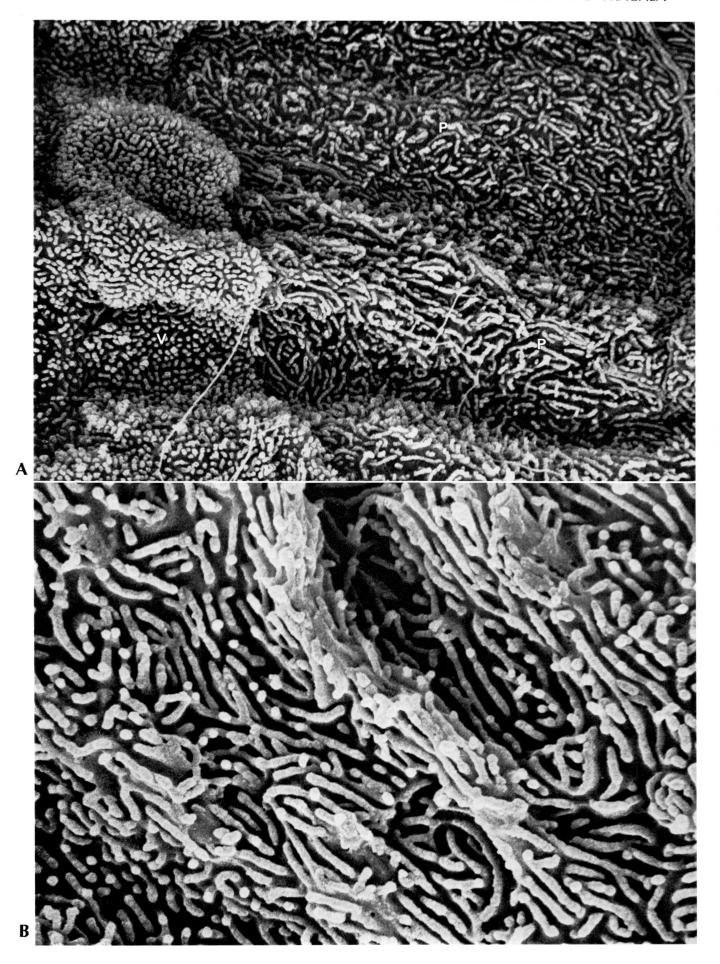

Plate IV-3   Vocal Cord.   Dog.

The vocal cord of the dog is covered by stratified squamous epithelium on its upper surface. The subglottal region of the larynx is covered by ciliated cells intermingled with goblet cells.

A shows the transitional portion of both epithelial types.

The upper right half of the micrograph represents the squamous epithelium covering the edge of the cord. The funnel-shaped and plicated pits are the orifices of the laryngeal glands which occur also on the cord.

The lower left area with parallel grooves is the subglottal ciliated epithelium. A narrow zone labeled T shows thickened, polygonal cells, most of which apparently are degenerating cells presumably abandoned in this region of the larynx.

B shows the squamous epithelium covering the upper surface of the vocal cord. The surface fine structures conspicuously differ from cell to cell.

C is a closer view of the laryngeal gland orifice which is seen in the center of the left figure.

A: × 400,   B: × 2,200,   C: × 16,000

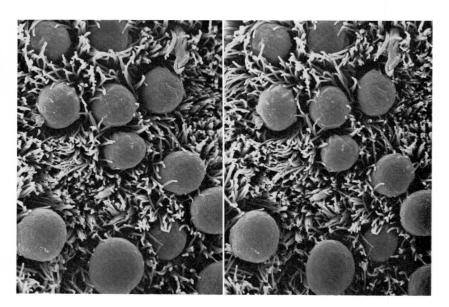

Fig. IV-1 Subglottal Region in Stereo. Dog.
The mucous epithelium of the laryngeal portion inferior to the vocal cord may contain numerous goblet cells. In this specimen the goblet cells appear conspicuous as the apical portion is swollen, apparently suggesting enhanced storage of mucous secretions.

× 2,700

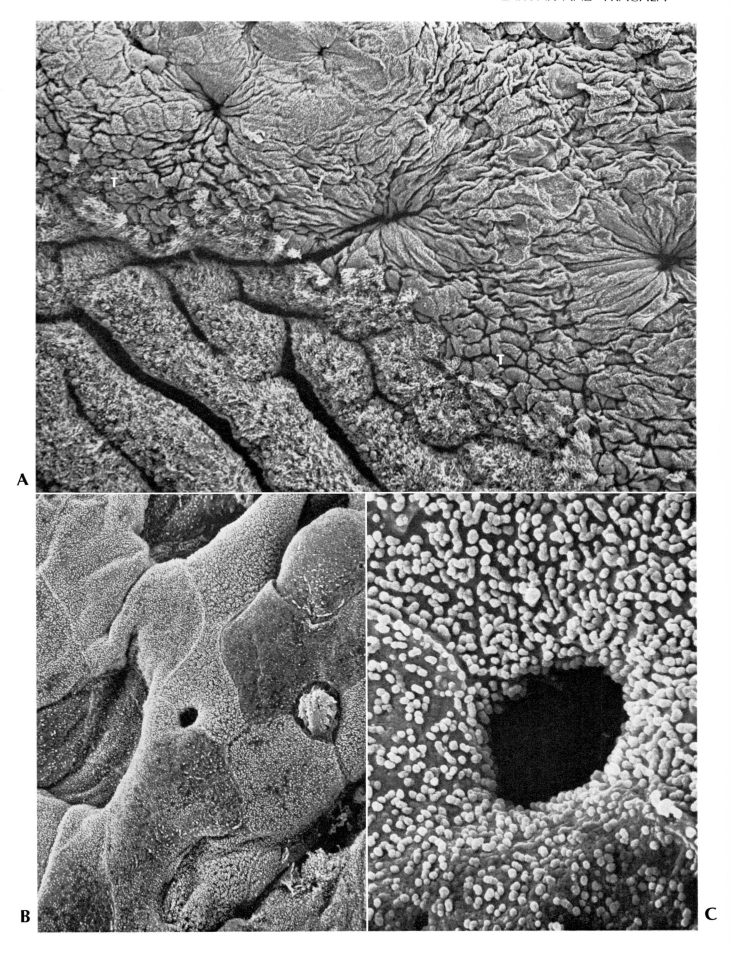

Plate IV-4   Subglottal Region of the Larynx.   Human.

The stratified squamous epithelium covering the vocal cord changes into pseudostratified non-ciliated epithelium a short distance inferior to the vocal edge. A certain distance more inferior, this epithelium is, in its turn, replaced by the ciliated epithelium of the trachea. The non-ciliated cells, therefore, are usually called transitional cells by otolaryngologists.

A shows the rather abrupt boundary between the transitional (non-ciliated) cell region and the ciliated surface of the trachea. Arrows indicate the orifices of laryngeal glands.

B demonstrates the typical "transitional cells" with their hexagonal apical face covered by short microvilli which are rather irregular in size and variably swollen at the tip.

C shows the boundary between the laryngeal and tracheal epithelium. Ciliated cells of the trachea are intercalated between the "transitional" cells.

A: X 780,   B: X 3,300,   C: X 3,600

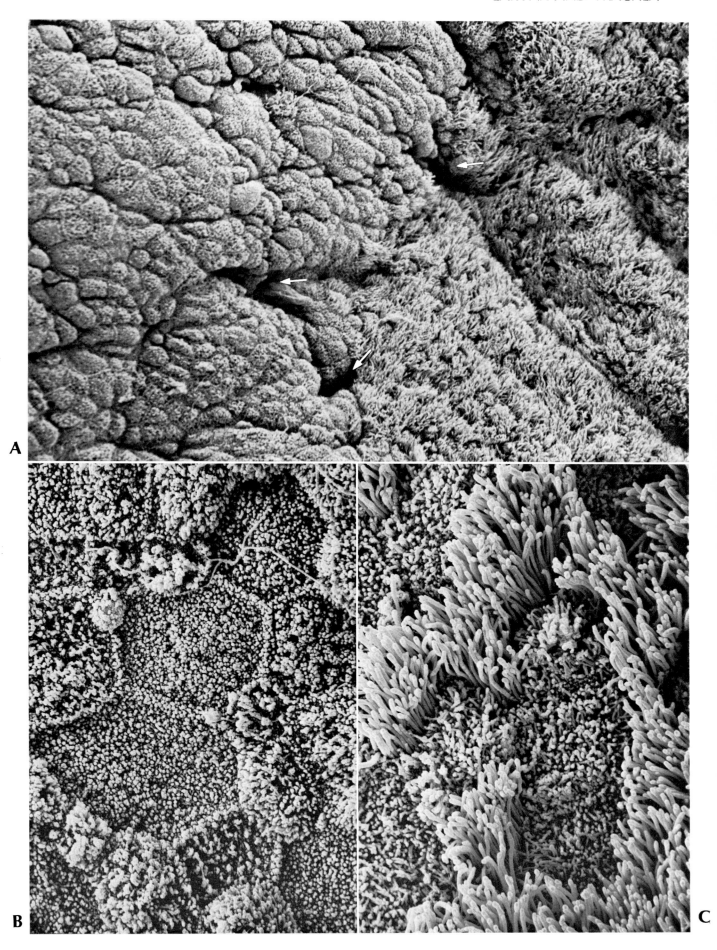

Plate IV-5   Trachea, the Uppermost Portion.   Human.

A. The gently undulating surface of the trachea is covered by a ciliated epithelium. As the ciliated cells essentially exclusively represent the surface cells and as long cilia grow very densely, the cell boundaries are not recognizable in the surface SEM view.

B shows a closer view of tracheal cilia, which are smoothly curved columns with a rounded end. A constriction occurs in many cilia in this micrograph shortly proximal to their end thus producing a short, rounded segment at the cilial tip. It is unknown whether this figure is an artifact caused by the specimen preparation or represents a natural phenomenon, such as the shedding of the distal portion of the cilia.

A: X 3,300,   B: X 16,000

(Specimen provided by Prof. M. Ohyama, Department of Otolaryngology, Kagoshima University School of Medicine).

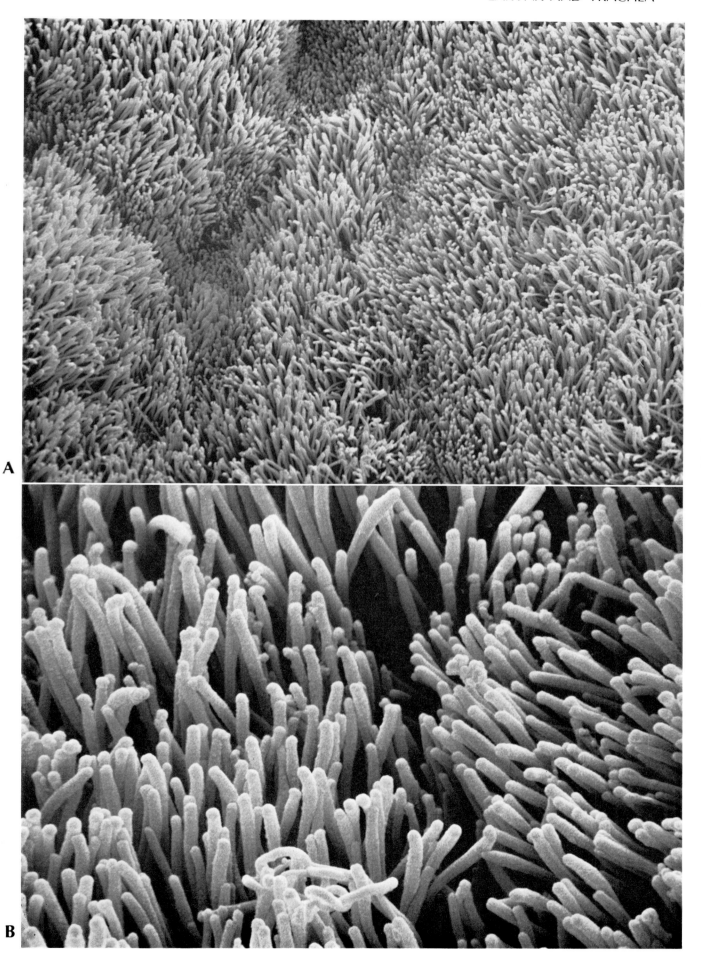

A

B

Lung

Plate IV-6   From Bronchiole to Alveoli.   Rabbit Lung.

In this lung fixed in a state of moderate inspiration, the air routes from the bronchioles to individual alveoli, where the gas exchange takes place, are demonstrated.

A *terminal bronchiole* (TB), accompanied by a branch of the pulmonary artery (PA), is shown in the upper right. It is divided into two short portions called *respiratory bronchioles* (RB). Several alveoli, which appear as dark round grooves in this micrograph, directly open to this portion.

The respiratory bronchiole further divides into *alveolar ducts* (AD). The alveolar duct represents a corridor surrounded by the open rooms of alveoli. The end portion of this corridor forms an antrum common to many alveoli, which is called the *alveolar sac* (AS); this may show a cloverleaf image in fractured tissue.

X 200

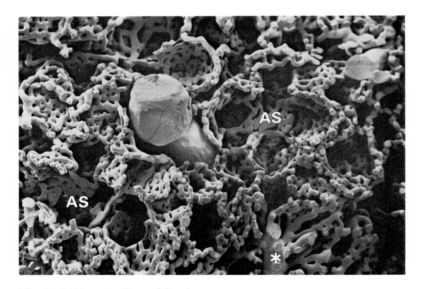

Fig. IV-2 Vascular Cast of Rat Lung.
The basket-like distribution of blood capillaries surrounding each alveolus is evident in this fracture view of a methacryl resin cast. The cloverleaf figures labeled AS seem to represent alveolar sacs. The connection of the capillaries with a probable arteriole (*) is seen in the lower right.

X 100

(Courtesy of Dr. T. Murakami, Department of Anatomy, Okayama University School of Medicine).

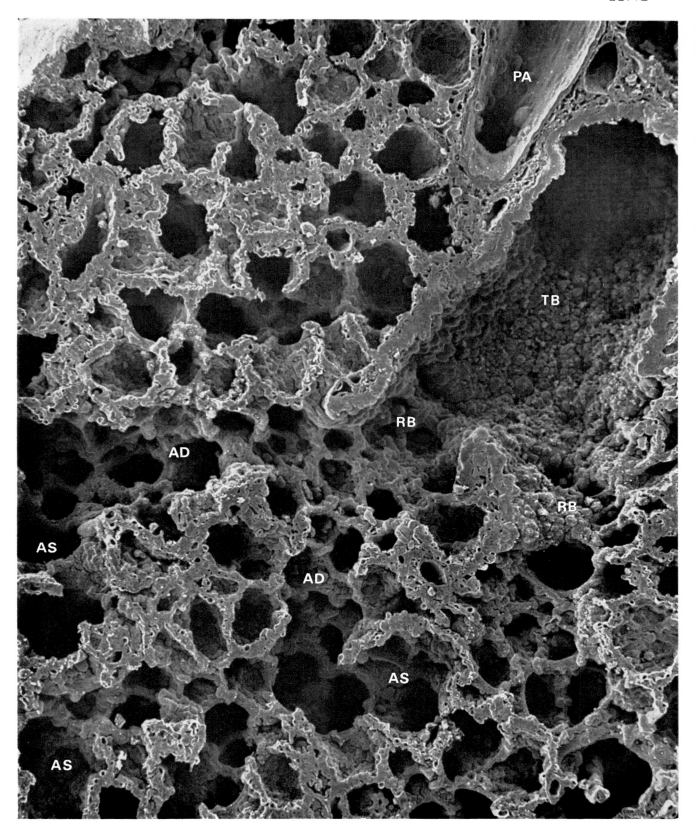

Plate IV-7   Clara Cells of the Terminal Bronchiole.   Rabbit.

Clara cells, first described in 1937 by the German anatomist Max Clara, are distributed at the end portion of the terminal bronchiole, where the ciliated epithelium rapidly decreases in height. The Clara cells are non-ciliated and higher than the ciliated cells which are distributed intermingled with them. The apical surface of the cell is smooth and strongly convex. The Clara cells are numerous and conspicuous especially in the rabbit.

TEM studies have shown that the Clara cells contain large amounts of smooth endoplasmic reticulum, and it has been proposed that the cells may produce lipidic secretions which may represent a sort of surfactant.

The *ciliated cells* in this micrograph possess rather smaller numbers of cilia with characteristically tapered ends. Also the cells are densely covered by microvilli.

   X 4,000

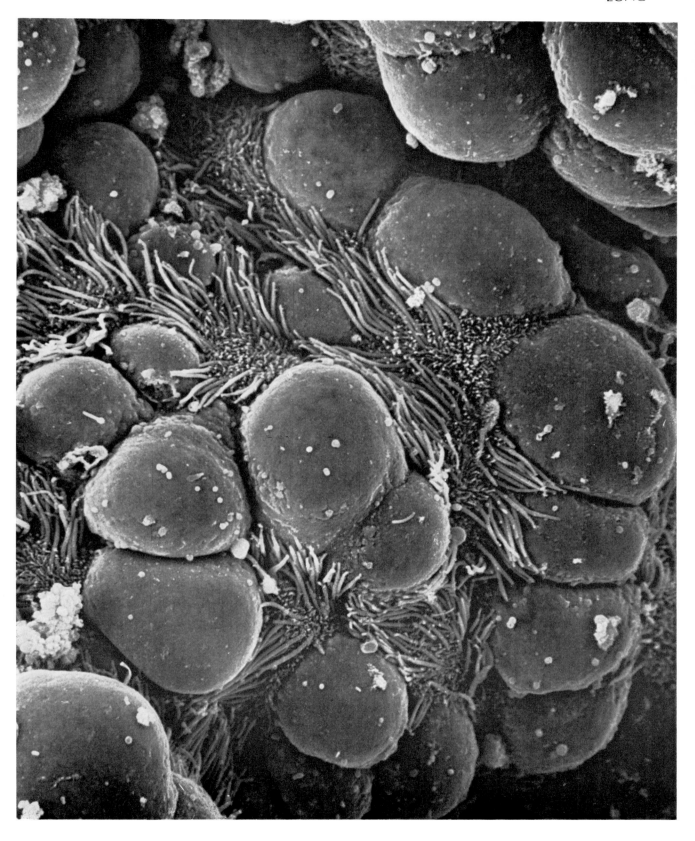

Plate IV-8   Blood - Air Barrier.   Rabbit Lung.

In this specimen the blood vessels are expanded considerably by vascular perfusion, and it is stressed that the pulmonary parenchyme comprises a complex of two labyrinthine spaces: a space for *blood* (B) and one for *air* (A). A very thin plate separates them, forming the *blood-air barrier* (arrows). This plate is conspicuously thickened at a few sites by the occurrence of great alveolar cells (G).

A further shows an alveolar macrophage (M) and, at the bottom, a bronchiole whose wall is undulated because of periodically arranged cartilage.

B is a closer view of the *blood-air barrier* (arrows), which is formed by an attenuated cytoplasm of squamous alveolar cells, that of endothelial cells and a thin basement membrane sandwiched by them. Nuclei of the squamous alveolar cells are labeled S, while those of the endothelial cells E. The endothelial surface is smooth with a few microvillous projections. The capillaries of this micrograph contain an erythrocyte and two lymphocytes.

A: X 2,400,   B: X 6,200

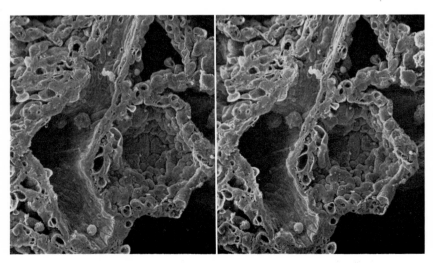

Fig. IV-3 Alveolar Wall in Stereo. Rabbit.
A bifurcating branch of the pulmonary vein is juxtaposed with alveoli. The fracture surface reveals numerous lumina of alveolar capillaries, which, in flat view, wind and swell into the alveolar lumen.

X 330

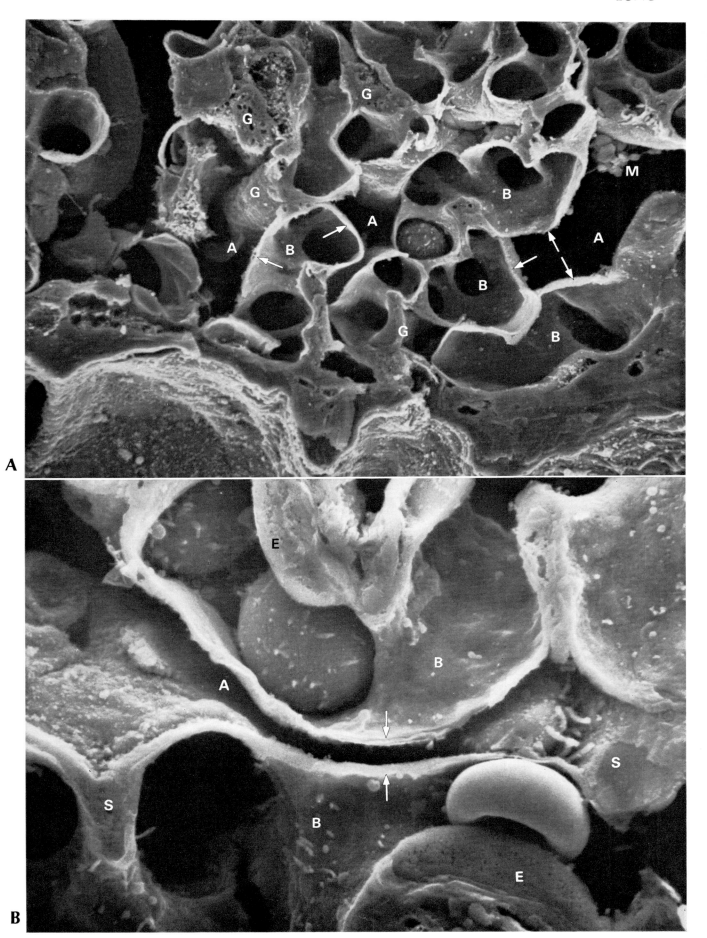

Plate IV-9   Great Alveolar Cell.   Rabbit.

The *great* or *type II alveolar cell* is known to produce dipalmityl lecithin and to store this substance in large, lysosome-like bodies called *lamellar bodies*. Previous TEM studies have revealed that the contents of the lamellar bodies are released by exocytosis, i.e., by opening of the sac of the body to the cell surface. Dipalmityl lecithin thus released is a potent surfactant on the air surface of alveoli protecting them from formidable collapse.

A demonstrates a great alveolar cell revealing both its fractured interior and surface structure. Some lamellar bodies are opened (arrows). The surface of the cell is covered by numerous, short microvilli, which are more clearly seen in the next micrograph. An alveolar macrophage (M) reveals its vacuole-rich cytoplasm.

B is the surface view of a great alveolar cell protruding into the alveolus. Short microvilli unevenly cover the cell. Grooves on the cell surface may be seen which are margined by the microvilli. As these microvilli lean towards each other, zipper-like forms may be noticed. More conspicuous are round holes (*) margined by the microvilli. They apparently correspond to the opened sacs of lamellar bodies.

A: X 10,000,   B: X 16,000

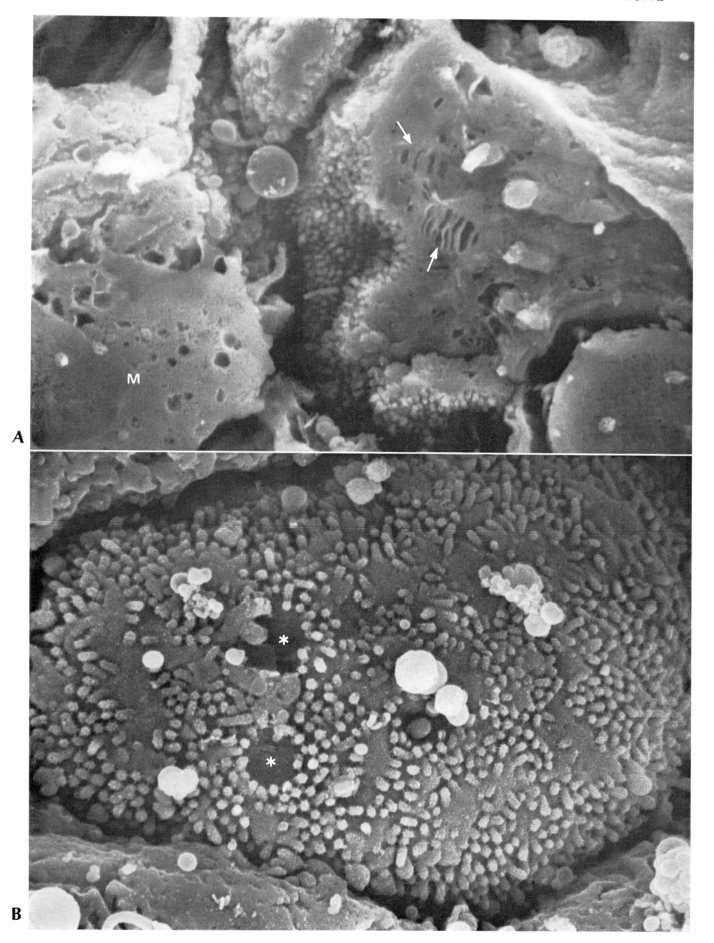

Plate IV-10

A. Alveolar Brush Cells. Bleomycin Administered Rat.

Brush cells characterized by a tuft of considerably thick, columnar microvilli are occasionally found dispersed in the alveolar epithelium. Hijiya (1978) revealed by SEM and TEM that the brush cells are increased in number after treatment of the animals with bleomycin. This micrograph is from the lung of a rat given intraperitoneal injections (daily 10 mg/kg) of bleomycin sulfate for one month.

The brush cells protrude their apical brush into the alveolar lumen but their cell body is covered by a thin cytoplasm of squamous alveolar cells, which, in this specimen, is partly eroded.

The nature of the brush cell is unknown, though its possible receptor function is suggested. It is also unknown whether the alveolar brush cell is identical with the cell of the same name in the trachea (see page 163).

X 12,000

B. Alveolar Macrophage. Rat Lung.

The *alveolar macrophage* plays an important role in the elimination of foreign bodies (dust particles, bacteria, etc.) in the alveoli. The origin of the cell has been disputed, but it is most generally believed to be derived from the tissue macrophage.

As macrophages generally do, the alveolar macrophage possesses numerous villous and bulbous microprojections.

X 18,500

(Plate IV-10A: Courtesy of Dr. K. Hijiya, Division of Surgery, Amagasaki Hospital, Amagasaki).

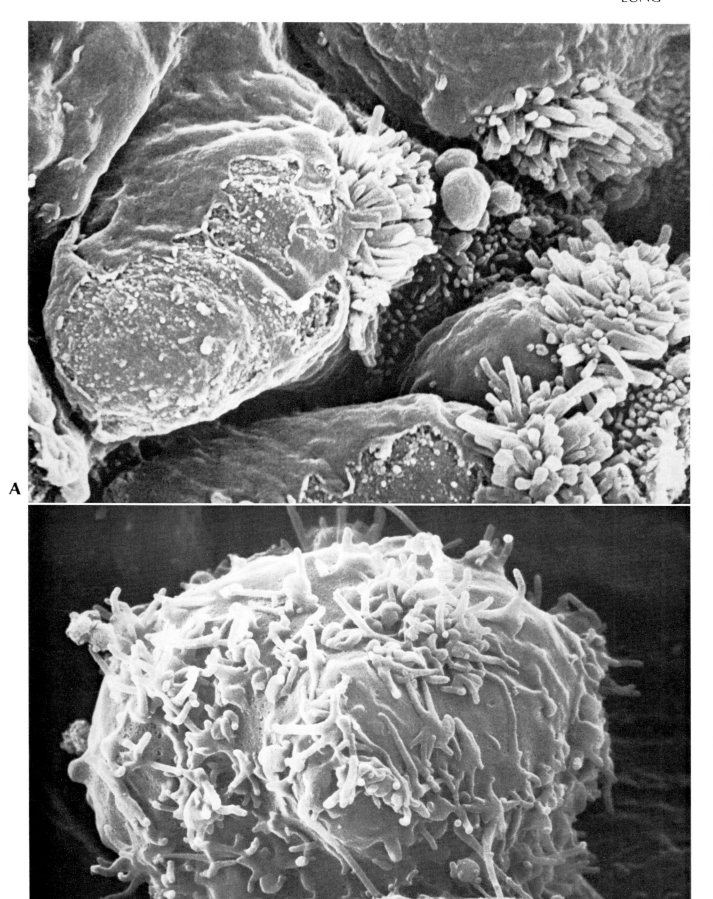

A

B

References

Andrews, P. M.: A scanning electron microscopic study of the extrapulmonary respiratory tract. *Amer. J. Anat.* **139**: 399—424 (1974).

Breipohl, W., C. Herberhold and R. Kerschek: Microridge cells in the larynx of the male white rat. Investigations by reflection scanning electron microscopy. *Arch. Oto-Rhino-Laryng.* **215**: 1—9 (1977).

Capella, C., E. Hage, E. Solcia and L. Usellini: Ultrastructural similarity of endocrine-like cells of the human lung and some related cells of the gut. *Cell Tiss. Res.* **186**: 25—37 (1978).

Castleman, W. L., D. L. Dungworth and W. S. Tyler: Intrapulmonary airway morphology in three species of monkeys: A correlated scanning and transmission electron microscopic study. *Amer. J. Anat.* **142**: 107—122 (1975).

Cutz, E., W. Chan, V. Wong and P. E. Conen: Endocrine cells in rat fetal lungs. Ultrastructural and histochemical study. *Lab. Invest.* **30**: 458—464 (1974).

Cutz, E., W. Chan, V. Wong and P. E. Conen: Ultrastructure and fluorescence histochemistry of endocrine (APUD-type) cells in tracheal mucosa of human and various animal species. *Cell Tiss. Res.* **158**: 425—437 (1975).

Greenwood, M. F. and P. Holland: The mammalian respiratory tract surface: A scanning electron microscopic study. *Lab. Invest.* **27**: 296—304 (1972).

Hijiya, K.: Ultrastructural study of lung injury induced by bleomycin sulfate in rats. *J. clin. Electron Microsc.* **11**: 245—292 (1978).

Ishii, N.: Electron microscopic studies on the respiratory system with relation to its aging — with special reference to the scanning electron microscopic observation. (In Japanese) *Nichidai Igaku Zasshi (Tokyo)* **36**: 959—971 (1977).

Krause, W. J. and C. R. Leeson: The postnatal development of the respiratory system of the opossum. I. Light and scanning electron microscopy. *Amer. J. Anat.* **137**: 337—356 (1973).

Kuhn III, C. and E. H. Finke: The topography of the pulmonary alveolus: Scanning electron microscopy using different fixations. *J. Ultrastr. Res.* **38**: 161—173 (1972).

Kuhn III, C. and F. Tavassoli: The scanning electron microscopy of elastase-induced emphysema. A comparison with emphysema in man. *Lab. Invest.* **34**: 2—9 (1976).

Lauweryns, J. M. and M. Cokelaere: Hypoxia-sensitive neuroepithelial bodies. Intrapulmonary secretory neuroreceptors, modulated by the CNS. *Z. Zellforsch.* **145**: 521—540 (1973).

Luciano, L., E. Reale and H. Ruska: Über eine "chemorezeptive" Sinneszelle in der Trachea der Ratte. *Z. Zellforsch.* **85**: 350—375 (1968).

Mellick, P. W., D. L. Dungworth, L. W. Schwartz and W. S. Tyler: Short term morphologic effects of high ambient levels of ozone on lungs of rhesus monkeys. *Lab. Invest.* **36**: 82—90 (1977).

Nowell, J. A., J. Pangborn and W. S. Tyler: Scanning electron microscopy of the avian lung. In: (ed. by) O. Johari: Scanning Electron Microscopy/1970. IIT Research Institute, Chicago, 1970 (p. 249—256).

Nowell, J. A. and W. S. Tyler: Scanning electron microscopy of the surface morphology of mammalian lungs. *Amer. Respirat. Dis.* **103**: 313—328 (1971).

Ogata, K.: Scanning electron microscopical studies on experimental mycoplasma pulmonis infection in tracheo-broncheal epithelium (included normal structure of tracheo-broncheal system). *J. clin. Electron Microsc.* **8**: 239—258 (1975).

Ohyama, M., Y. Miyoshi, S. Yamamoto, T. Taniguchi, T. Fujita and K. Adachi: Surface ultrastructure of pathological mucosa of the human larynx. Special reference to cinematographic visualization of scanning electron microscopic pathology following radiotherapy in the laryngeal cancers. *Otologia* **23**: 646—659 (1977a).

Ohyama, M., S. Yamamoto, T. Fujita and K. Adachi: An experimental study of influence on surface architecture of respiratory mucosa after tracheal fenestration. Ultrastructural analysis with the aid of a few method, "scanning electron microscopic cinematography". *Otologia* **23**: 666—686 (1977b).

Okada, Y.: The ultrastructure of the Clara cell in the bronchiolar epithelium. *Bull. Chest Dis. Res. Inst., Kyoto Univ.* **3**: 1—10 (1969).

Rhodin, J. and T. Dalhamn: Electron microscopy of the tracheal ciliated mucosa in rat. *Z. Zellforsch.* **44**: 345—412 (1956).

Smolich, J. J., B. F. Stratford, J. E. Maloney and B. C. Ritchie: Postnatal development of the epithelium of larynx and trachea in the rat: scanning electron microscopy. *J. Anat.* **124**: 657—673 (1977).

Taira, K. and S. Shibasaki: A fine structure study of the non-ciliated cells in the mouse tracheal epithelium with special reference to the relation of "brush cells" and "endocrine cells". *Arch. histol. jap.* **41**: 351—366 (1978).

Tateishi, R.: Distribution of argyrophil cells in adult human lungs. *Arch. Pathol.* **96**: 198—202 (1973).

Tyler, W. S., A. A. de Lorimier, A. G. Manus and J. A. Nowell: Surface morphology of hypoplastic and normal lungs from newborn lambs. In: (ed. by) O. Johari and I. Corvin: Scanning Electron Microscopy/1971. IIT Research Institute, Chicago, 1971 (p.305—312).

CHAPTER V

URINARY SYSTEM

Kidney

The early SEM studies on the kidney were focused on the structures of the *glomeruli* (Buss and Krönert, 1969; Arakawa, 1970; Fujita *et al.*, 1970; Spinelli *et al.*, 1972; Andrews and Porter, 1974; Arakawa and Tokunaga, 1974; Bulger *et al.*, 1974). The entire body of the glomerular *podocyte* with its fantastically delicate processes could thus be visualized. SEM observations have elucidated that the end-feet of the fern-like podocyte processes are interdigitated always with the processes from different cells, and an end-foot (pedicle) never is juxtaposed with one from the same podocyte (Fujita *et al.*, 1970; Miyoshi *et al.*, 1971).

Recent TEM studies have suggested that the podocytes are contractile in function. The elaborately interdigitated processes, thus, are presumed to control the caliber of the glomerular capillaries on one hand, and, on the other hand, the size of the interpedicular slits through which the primary urine is filtrated.

The main element of the glomerular filtration membrane is the *basement membrane* on which the podocyte pedicles lie. The other side of the basement membrane is lined by *a pored endothelium*, whose SEM view has been also analyzed (Fujita *et al.*, 1976).

The *renal tubules* are involved in the reabsorption of urine and other multiple functions and different portions of the tubule are lined by different types of cells. Although successful SEM observation of the renal tubule has been retarded compared with that of the glomerulus, excellent papers have been published in the last few years adding some knowledge to the previous TEM observations of the tubule. Cells with brush border microvilli, single cilia and wart-like or ridge-like microprojections are demonstrated under the SEM, revealing their precise shapes and distribution (Andrews and Porter, 1974; Bulger *et al.*, 1974; Andrews, 1975, Hücker *et al.*, 1975).

Stereo-fine structures of different portions of the nephron will be demonstrated in this chapter, from the glomerulus down to the distal tubule.

Plate V-1　Surface View of Glomerulus.　Rat.

Bowman's capsule is opened and a glomerulus is exposed. The latter consists of winding loops of blood capillaries covered, from the outside, by *podocytes* extending their processes. The thickened nuclear portion of the podocytes (P) and its processes shaped like fern leaves are evident at this low magnification.

Bowman's capsule consists of a thin epithelium supported by a basement membrane and collagen fibers which are visible in the lower part of this micrograph. The epithelial cells possess *single cilia* (arrows).

　X 2,000

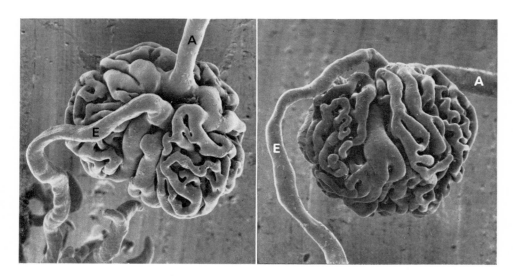

Fig. V-1　Vascular Cast of Glomerulus.　Human Newborn.
The same cast is seen from the side of the vascular pole (left) and laterally (right). It is clear from these micrographs that the glomerular vasculature is divided into several *lobes*, each of which is composed of a complicated network of winding capillaries.

A indicates the vas afferens, whereas E shows a vas efferens which is branched to supply the renal tubules as the micrograph on the left shows.

　X 330

(Courtesy of Dr. T. Murakami, Department of Anatomy, Okayama University Medical School).

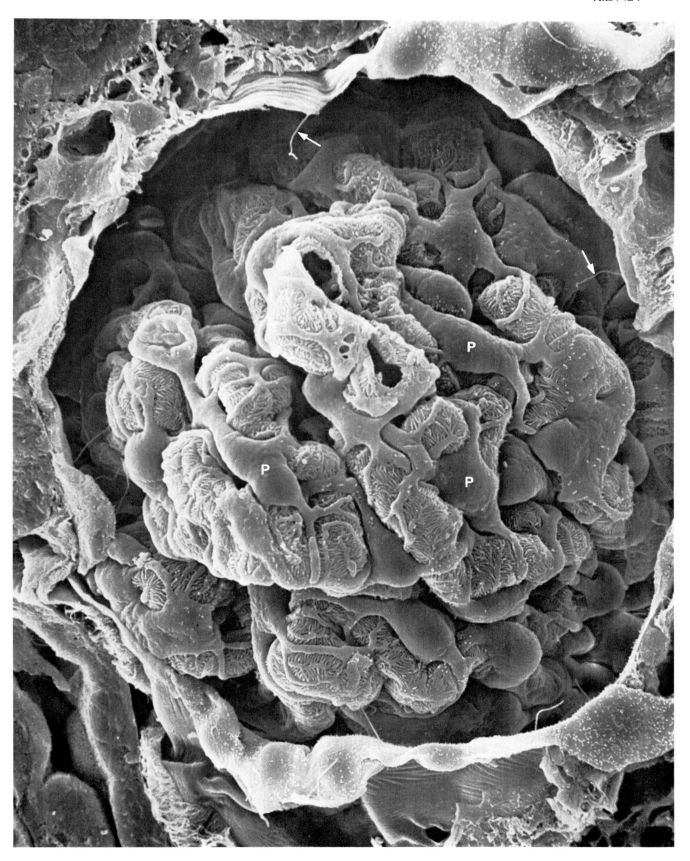

Plate V-2 Fractured Glomerulus. Rat.

A glomerulus ensheathed by Bowman's capsule is fractured approximately along its equatorial plane. *Glomerular capillaries* are cut in various directions. The nuclear portions of endothelial cells occasionally bulge into the capillary lumen. On the outside of the capillary wall the rounded cell bodies of podocytes (P) and their processes are attached.

This picture demonstrates that the tortuous capillaries are connected by *mesangium* just like the intestines are by mesenterium and they are grouped into several lobules which are evident also in Plate V-1. The large arrows indicate the narrow but deep spaces separating the lobules.

Bowman's capsule (B) consists of a thin epithelium sheet showing slight swellings of nuclei (small arrows). Outside of the capsule are seen sections of proximal (PC) and distal convolutions (DC) and some blood capillaries (C).

X 1,400

(Reproduced from T. Fujita, J. Tokunaga and M. Edanaga: *Cell Tiss. Res.* 166: 299—314, 1976).

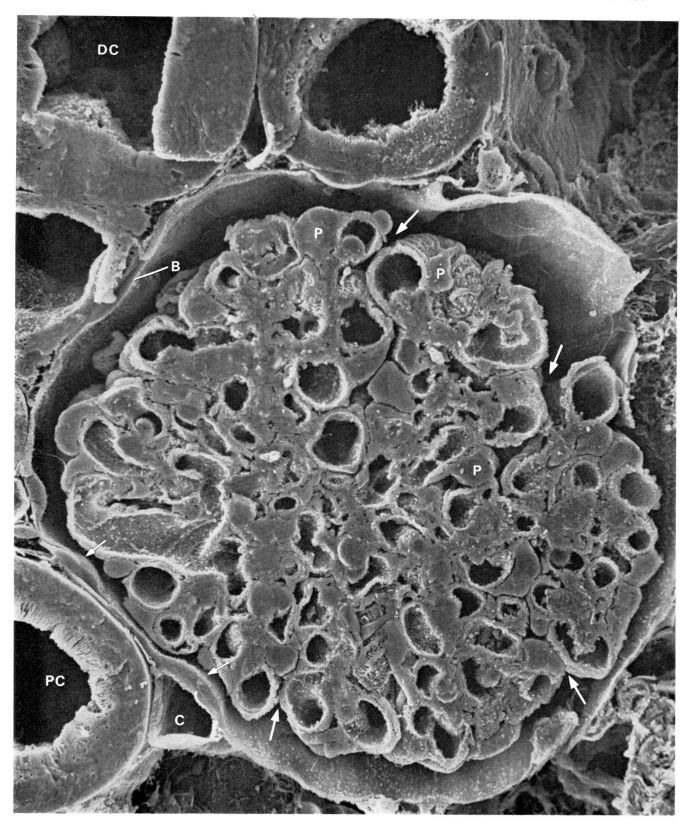

Plate V-3 Podocytes of Glomerulus. Rat.

This micrograph is a closer view of a part of Plate V-1, and shows how podocytes cover the wall of the glomerular capillaries. The nuclear portion of the podocytes issues three to five thick processes (primary processes) which branch into secondary processes. These, either after dividing into smaller branches or directly, from the axis from which slender end-feet (pedicles) are issued bilaterally. The axial process and its end feet thus form a "fern leaf" attached to the capillary wall.

The end-feet of a fern leaf derived from a podocyte are alternately interdigitated with those from other cells. It is thus a strict rule that end-feet *only from different cells*, can be juxtaposed.

The podocytes show some granular and short villous microprojections. Single cilia are lacking in this species.

At the top of the micrograph Bowman's capsule is seen with its cilia.

X 5,000

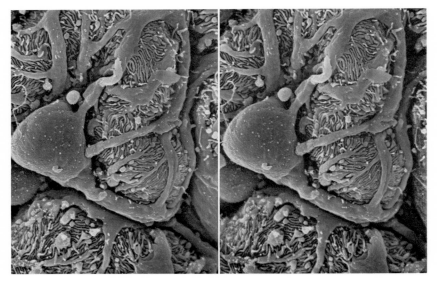

Fig. V-2 Podocyte in Stereo. Rabbit.
In the rabbit the processes of the podocytes are somewhat less regular in shape, size and branching pattern than in the rat, but the principle of interdigitation of their end feet is the same.

X 2,600

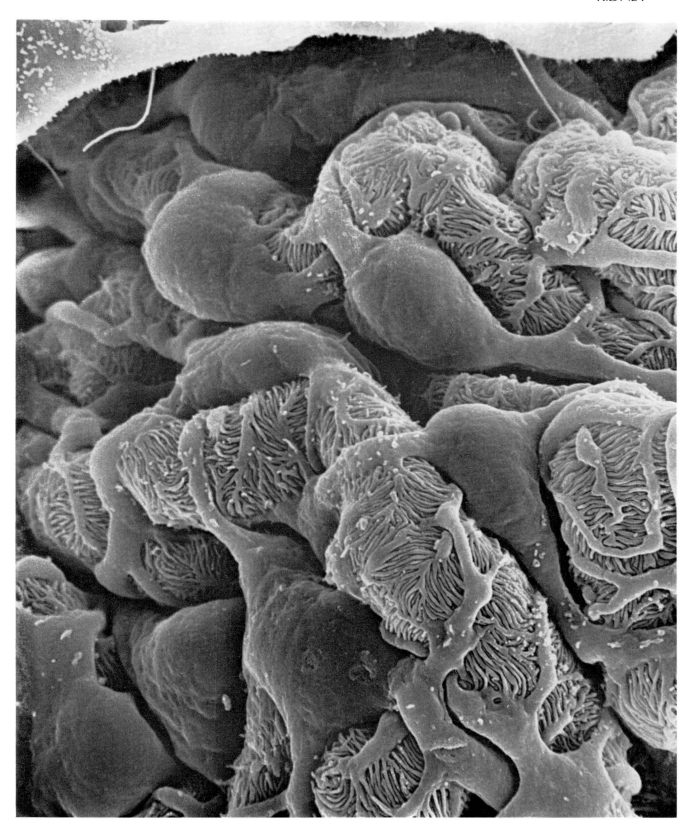

Plate V-4 Glomerular Filtration Membrane. Rat.

This micrograph shows a grazing fracture of the glomerular filtration membrane seen from the urinary side. The *basement membrane* (B) which represents the main filter for producing the primary urine appears layered in structure.

Podocyte pedicles alternately interdigitated with those from different cells are attached to the basement membrane, and after they are removed they may leave "foot-marks" (arrows) on the basement membrane. The spaces between the pedicles show some coagulated matter which may partly correspond to the polysaccharide-rich substances involved in urine filtration.

The hemal portion of the filtration membrane is the pored cytoplasm of an *endothelial cell* (E), which is seen from the back (urinary) side in this micrograph, though partly destroyed.

X 20,000

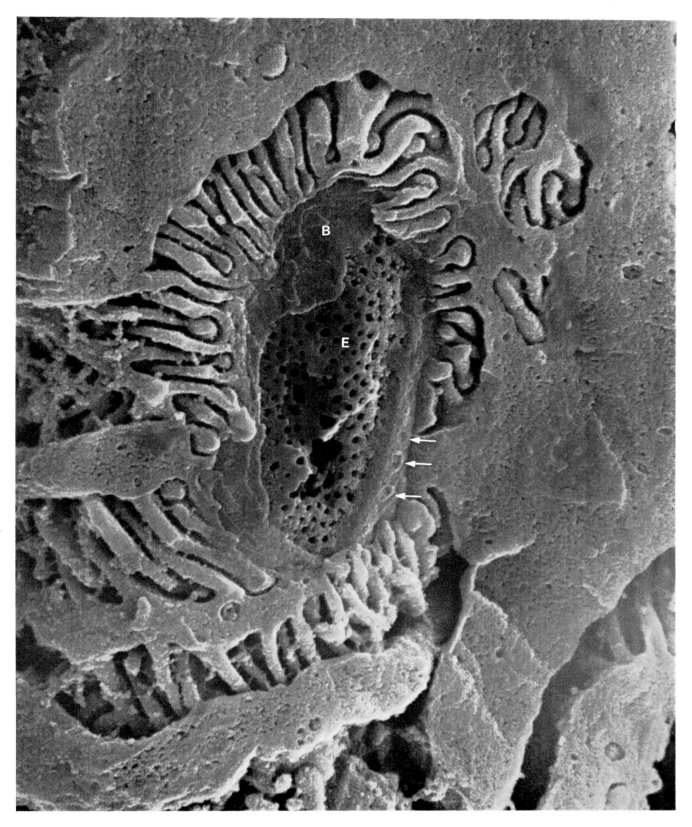

Plate V-5 Glomerular Capillary. Rat.

In this micrograph which is a closer view of a portion in Plate V-2 a glomerular capillary is opened longitudinally, revealing the profile of its wall or the filtration membrane and a luminal view of the endothelial lining at the same time.

The endothelial sheet is divided into pored compartments called *areolae fenestratae* (Fujita *et al.*, 1976) by *cytoplasmic crests* radiating from the nuclear swelling (not shown in this micrograph) of the endothelial cell. A crest also occurs along the cell margin so that a slit-like cell boundary is found between the marginal crests of adjacent cells (arrows). Some irregular-shaped elevations and villous microprojections also are seen on the endothelial surface.

X 14,000

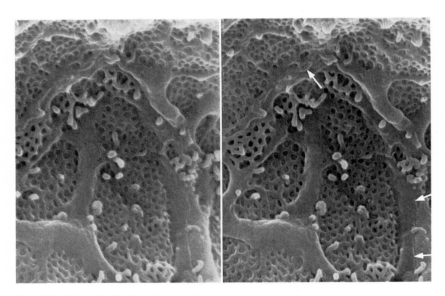

Fig. V-3 Endothelial Surface in Stereo. Rat Glomerulus.
The *cytoplasmic crests* and the *areolae fenestratae* are evident. Gradations from microvilli to pored sheet are demonstrated. A part of the pored sheet apparently is a newly built one above the level of the old one. Such an image thus is interpreted as showing the process of renewal of the areola fenestrata. Arrows indicate cell boundaries.

X 13,000

(Reproduced from T. Fujita *et al.*: *Cell Tiss. Res.* **166**: 299–314, 1976).

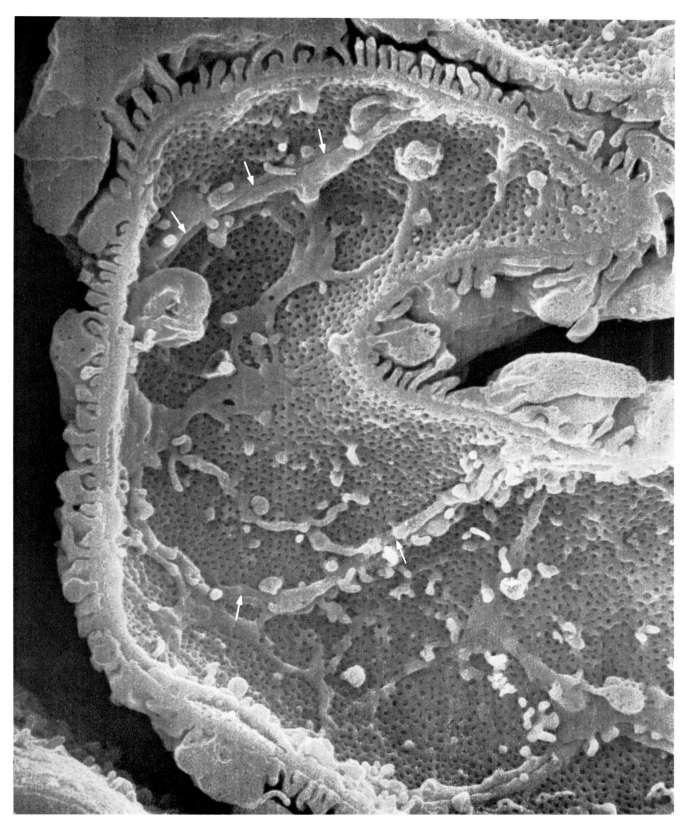

Plate V-6
A. High Power View of Glomerular Endothelium. Rat.

The *basement membrane* is sandwiched by dentate profiles of podocyte end-feet and by pored endothelium.

In the *endothelium* cytoplasmic crests and areolae fenestratae are evident. A cell boundary runs between *marginal crests* of adjacent cells. What SEM study recently revealed is that the *endothelial pores* are much more irregular in size (30—150 nm) than previously believed on the basis of TEM observations. The pores apparently lack in diaphragm as is known from TEM studies in this species (Latta, 1973), and the flat bottom of the pores corresponds to the basement membrane. A knob-like swelling may be seen in some pores.

X 33,000

B. Urinary Pole. Rat.

Bowman's capsule is lined by thin epithelial cells studded with granular microprojections and possessing a single cilium. This structure abruptly is changed into the brush-bordered epithelium of *proximal convolution* at the connection of the capsule and the tubule (urinary pole).

The micrograph shows the orifice of a dark and tortuous tunnel through which the urine is to be transferred and reabsorbed.

X 4,600

(Plate V-6: Reproduced from T. Fujita *et al.*: *Cell Tiss. Res.* **166**: 299—314, 1976).

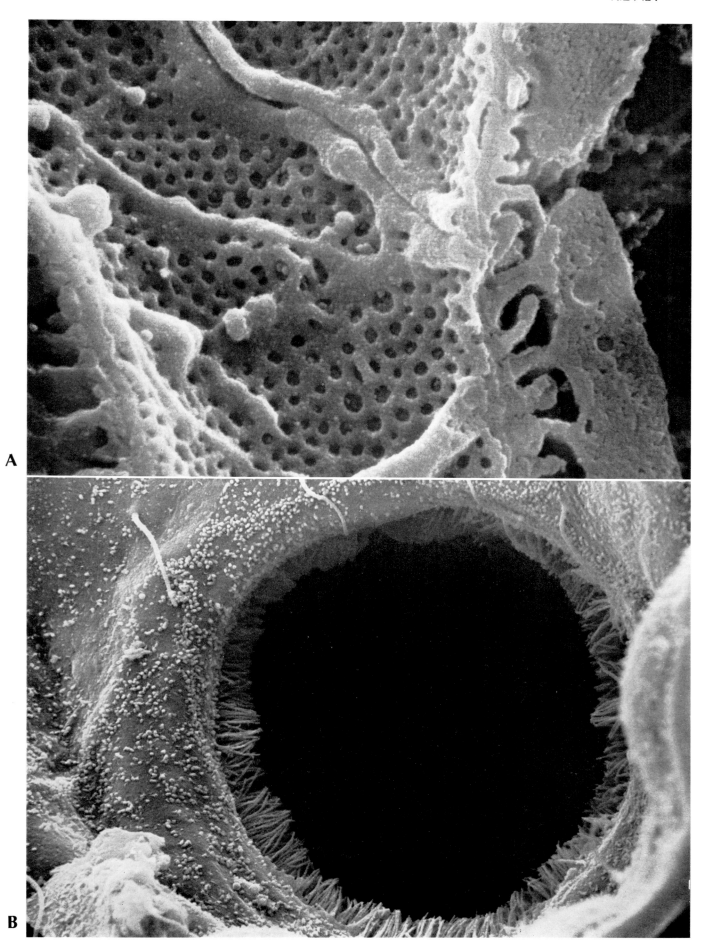

A

B

Plate V-7

A. Proximal Convolution. Rat.

A portion of the proximal convolution is fractured longitudinally. The epithelial cells are densely covered by microvilli corresponding to the *brush border*. The basal portion of the epithelium shows a complicated profile of *cell interdigitation*. These structures of the cells are involved in the active reabsorption of urine in this portion of the urinary tubule.

A *macrophage* (M) is shown in the lumen of the tubule. The microvilli covering this cell are conspicuously different in shape, size and coating substance from those of the tubule epithelium. SEM observations indicate that macrophages are more numerous in the lumen and wall of the proximal convolution than believed previously. The significance of their phagocytotic activity in this site should be re-evaluated.

N: nucleus of the epithelial cell.

X 11,000

B. Brush Border of Proximal Convolution. Rat.

In this micrograph the *microvilli* of the brush border are slender columns of uniform length and thickness with a small knob at their ends. Occasional long and thicker strings are *single cilia*. *Cony pits* are frequently found in the brush border (Tokunaga *et al.*, 1974; Andrews, 1975); their significance is unknown. Small balls of possible secretional nature may be seen in the pit as shown in this micrograph (arrows).

Vesicles of different sizes are seen in the apical cytoplasm of the epithelial cell. They are mostly pinocytotic vesicles and some of them are apparently continuous to the cell surface.

X 15,500

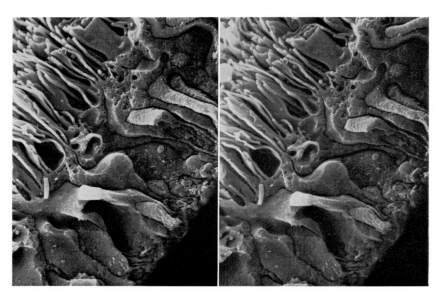

Fig. V-4 Proximal Tubule in Stereo. Rat.
In this specimen the fracture occurred mainly along the cell boundaries so that the elaborate interdigitation of adjacent epithelial cells is clearly demonstrated. Noteworthy also is the plate-like shape of the microvilli in this specimen.

X 6,700

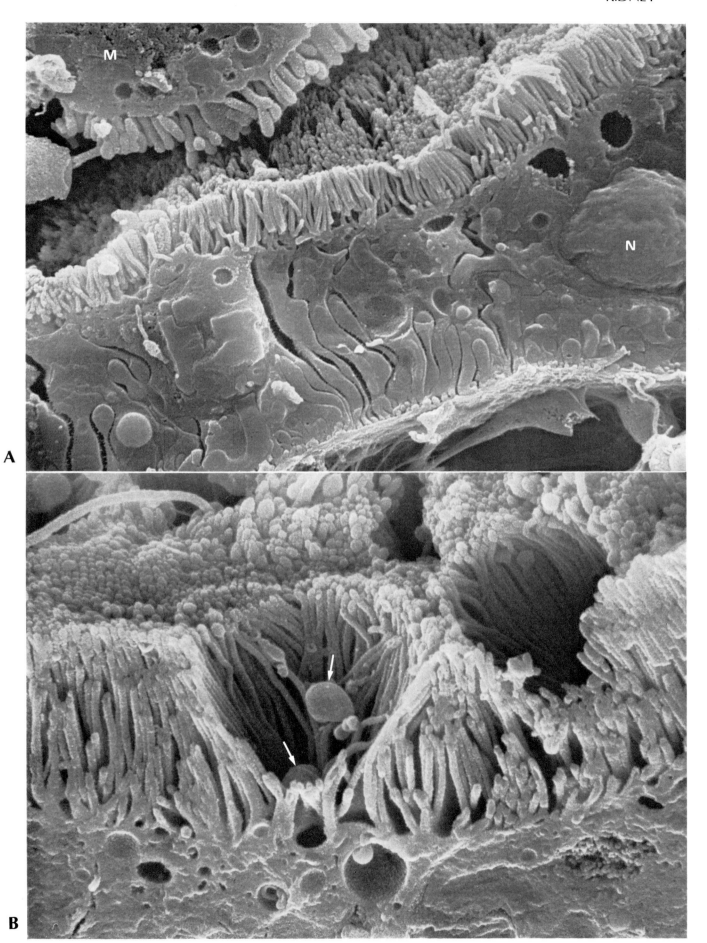

Plate V-8
A. Henle's Loop, Crosscut. Rabbit.

In a superficial portion of the renal medulla thick portions of Henle's loop are crosscut. Their epithelium is paved by cells which are polygonal plates of a moderate thickness. The nuclear portion is slightly thickened and possesses a *single cilium*. The cell surface is covered by granular and low villous microprojections.

At the upper left is seen a thin portion of Henle's loop which is formed by a very thin epithelium. The luminal surface of this portion is relatively smooth with some irregular microvilli in this species.

On the left hand a small part of a proximal tubule (straight portion) is seen.

The spaces left between the tubules are occupied mainly by *blood capillaries* (C) and partly by a reticular tissue containing process-rich fibroblasts (∗).

X 3,000

B. Thin Portion of Henle's Loop. Rat.

A thin portion of Henle's loop is longitudinally opened. The low epithelial cells are covered by very short microvilli. Most, if not all, of the cells possess a *single cilium*. Cell boundaries are not clear.

Although their distribution is variable among species, the occurrence of *single cilia* has been found on most epithelial cells lining the mammalian nephron. TEM analysis has revealed that their microtubules consist of 7—9 peripheral doublets and no central singlet filaments, suggesting that they may be only weakly motile causing local turbulence in the urinary flow. They may be chemosensory in function or possibly functionless (Flood and Totland, 1977).

X 6,000

C. Intercalated Cell in Distal Tubule. Rat.

In the distal and collecting tubules of man and rat, cells with a convex surface covered by microplicae and/or microvilli but lacking, in contrast to other cells, in cilium are intercalated (Andrews and Porter, 1974; Hücker *et al.*, 1975; Pfaller and Klima, 1976). These cells are believed to correspond to the *"dark"* or *"intercalated cells"* known in these tubules from TEM (Griffith *et al.*, 1968), but their function is unknown.

In the upper right a cilium of an ordinary cell is shown.

X 12,000

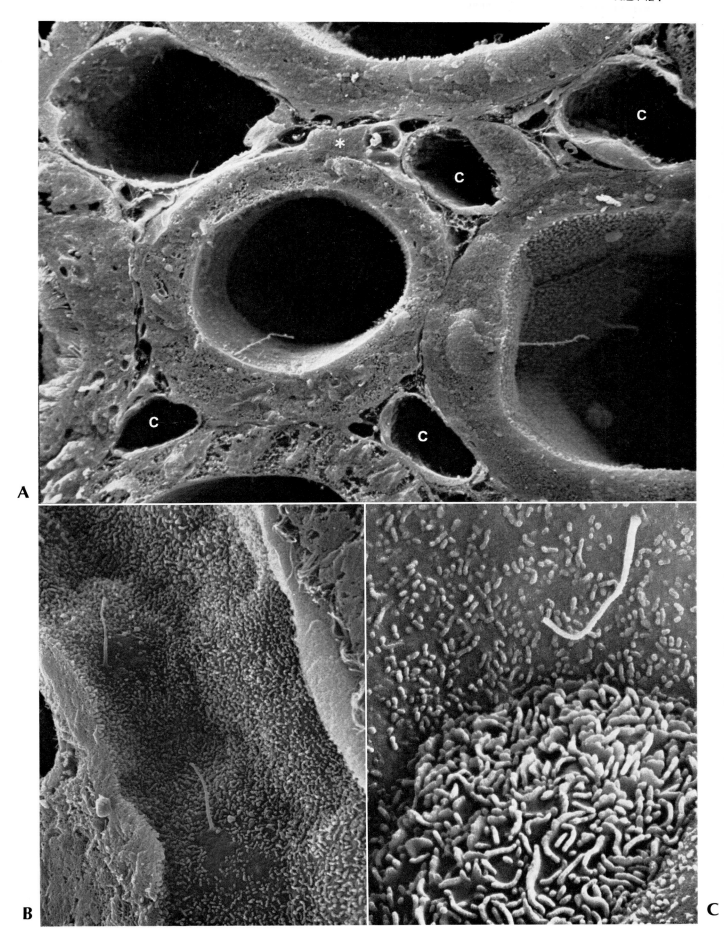

A

B

C

Ureter and Bladder

Plate V-9
A. Urinary Bladder. Rabbit.

The moderately extended mucous membrane of rabbit bladder was freeze-cracked so that the horizontally fractured surface as well as the natural luminal surface could be demonstrated.

The surface cells of the *transitional epithelium* are pentagonal or hexagonal in outline. They are markedly rough in surface because of the densely developed microridges of the cell. Conspicuous are the patches of *small cells* with a more rounded outline and smoother surface. In agreement with the view of Wong and Martin (1977), who in the guinea pig demonstrated small surface cells with the same characteristics by SEM, we conceive that they are young surface cells which have newly emerged from the intermediate layer.

X 320

B. Surface Cells of Bladder Epithelium. Rabbit.

The surface cells of the transitional epithelium of the bladder are provided with a network of *microridges,* whose appearance differs from cell to cell likely according to the maturation of the cell.

As known from TEM and chemical studies, the surface of these cells is covered by a specialized thickening of the plasma membrane and this contains keratin or a keratin-like scleroprotein which makes the bladder epithelium relatively impermeable to water and salts (Hicks, 1966). The SEM view shown here gives an impression of the unusually rigid nature of the cell surface.

Between large cells a very small cell area of triangular shape (∗) is intercalated. Granular microprojections of fairly uniform size are gathered along the cell boundaries (arrows).

X 7,800

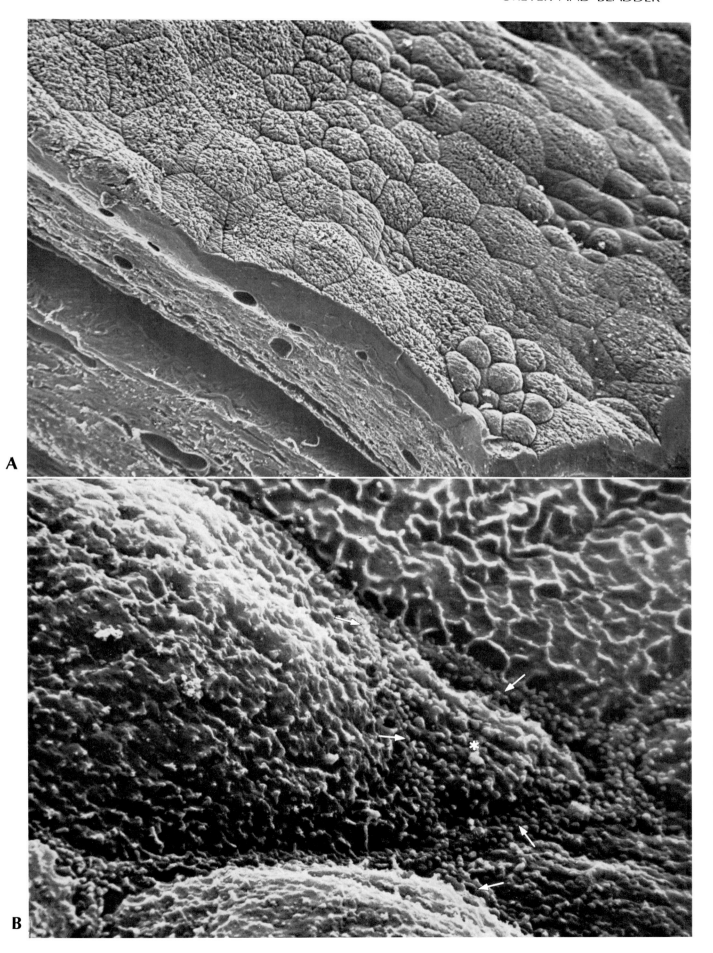

A

B

Plate V-10   Transitional Epithelium.   Pig Renal Calyx.

The transitional epithelium of the *renal calyx* of the pig was, before fixation, dipped in 2.5% EDTA for 2 hrs and then sprayed with physiological saline. By this method the epithelial cells have been partly removed.

A. The epithelial cells reveal their lateral surface. The *surface cells* with their swollen heads extend a slender process (P) to the basement membrane. *Intermediate* (M) and *basal cells* (B) all stand on the basement membrane with their more or less stout foot.

This SEM finding in the pig confirms the view reached by Tanaka (1962) in man using polarization microscopy and by Petry and Amon (1966) in various mammals using TEM that *all* the cells of the transitional epithelium reach the basement membrane. However, it has been claimed in some species (rat and shrew) that the surface cells are devoid of a basal process and separated from the epithelial base (Scheidegger and Ludwig, 1977).

B. In this micrograph three epithelial cells of the intermediate zone are left on the basement membrane, indicating their characteristic piriform shape with a *basal process.*

A: X 4,400,   B: X 7,400

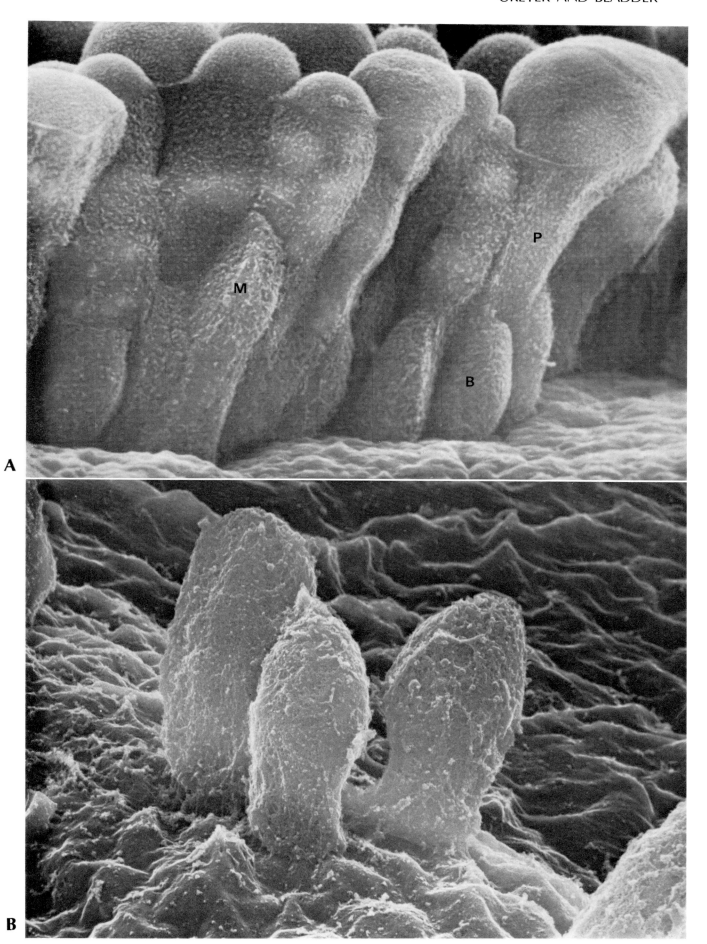

A

B

References

Andrews, P. M.: Scanning electron microscopy of human and rhesus monkey kidneys. *Lab. Invest.* **32**: 610–618 (1975).

Andrews, P. M. and K. R. Porter: A scanning electron microscopic study of the nephron. *Amer. J. Anat.* **140**: 81–116 (1974).

Arakawa, M.: A scanning electron microscopy of the glomerulus of normal and nephrotic rats. *Lab. Invest.* **23**: 489–496 (1970).

Arakawa, M. and J. Tokunaga: Further scanning electron microscope studies of the human glomerulus. *Lab. Invest.* **31**: 436–440 (1974).

Bulger, R. E., F. L. Siegel and R. Pendergrass: Scanning and transmission electron microscopy of the rat kidney. *Amer. J. Anat.* **139**: 483–502 (1974).

Buss, H. and W. Krönert: Zur Struktur des Nierenglomerulum der Ratte. Rasterelektronenmikroskopische Untersuchungen. *Virchows Arch. Abt. B Zellpathol.* **4**: 79–92 (1969).

Flood, P. R. and G. K. Totland: Substructure of solitary cilia in mouse kidney. *Cell Tiss. Res.* **183**: 281–290 (1977).

Fujita, T., J. Tokunaga and M. Edanaga: Scanning electron microscopy of the glomerular filtration membrane in the rat kidney. *Cell Tiss. Res.* **166**: 299–314 (1976).

Fujita, T., J. Tokunaga and M. Miyoshi: Scanning electron microscopy of the podocytes of renal glomerulus. *Arch. histol. jap.* **32**: 99–113 (1970).

Griffith, L. D., R. E. Bulger and B. F. Trump: Fine structure and staining of mucosubstances on "intercalated cells" from the rat distal convoluted tubule and collecting duct. *Anat. Rec.* **160**: 643–662 (1968).

Hicks, R. M.: The permeability of rat transitional epithelium. *J. Cell Biol.* **28**: 21–31 (1966).

Hücker, H., H. Frenzel and D. Skoluda: Scanning electron microscopy of the distal nephron and calyx of the human kidney. *Virchows Arch. B Cell Pathol.* **18**: 157–164 (1975).

Latta, H.: Ultrastructure of the glomerulus and juxtaglomerular apparatus. In: (ed. by) J. Orloff and R. W. Berliner: Handbook of physiology, Sect. 8: Renal physiology. American Physiological Society, Washington, 1973 (p.1–29).

Miyoshi, M., T. Fujita and J. Tokunaga: The differentiation of renal podocytes. A combined scanning and transmission electron microscope study in rats. *Arch. histol. jap.* **33**: 161–178 (1971).

Petry, G. and H. Amon: Licht- und elektronenmikroskopische Studien über Struktur und Dynamik des Übergangsepithels. *Z. Zellforsch.* **69**: 587–612 (1966).

Pfaller, W. and J. Klima: A critical reevaluation of the structure of the rat uriniferous tubule as revealed by scanning electron microscopy. *Cell Tiss. Res.* **166**: 91–100 (1976).

Scheidegger, G. and K. S. Ludwig: Elektronenmikroskopische Untersuchung am Übergangsepithel der Hausspitzmaus *(Crocidura russula)*. *Experientia* **33**: 1507–1509 (1977).

Spinelli, F., H. Wirz, C. Brücher and G. Pehling: Non existence of shunts between afferent and efferent arterioles of juxtamedullary glomeruli in dog and rat kidney. *Nephron* **9**: 123–131 (1972).

Tanaka, K.: Polarisationsoptische Analyse der Übergangsepithelien des Menschen. *Arch. histol. jap.* **22**: 229–236 (1962).

Tokunaga, J., M. Edanaga, T. Fujita and K. Adachi: Freeze cracking of scanning electron microscope specimens. A study of the kindney and spleen. *Arch. histol. jap.* **37**: 165–182 (1974).

Wong, Y. C. and B. F. Martin: A study by scanning electron microscopy of the bladder epithelium of the guinea pig. *Amer. J. Anat.* **150**: 237–246 (1977).

CHAPTER VI

MALE REPRODUCTIVE SYSTEM

SEM study of the reproductive system started with the observation of animal (Dott, 1969) and human (Fujita *et al.*, 1970; Lung and Bahr, 1972) *spermatozoa*, as these were free cells ready to be examined by the SEM if only the sperm plasma could be eliminated before fixation.

The study then proceeded towards the SEM observation of atypical spermatozoa (Fujita *et al.*, 1970; Lacy *et al.*, 1974; Fujita, 1975). After simplification and routinization of specimen preparation techniques, the SEM will be widely used by urologists in the near future to analyse the ejaculated spermatozoa of subfertile and infertile men. Besides this, SEM study has stimulated the advance of comparative morphology of spermatozoa (Matano *et al.*, 1976).

Furthermore, the eye of the SEM pursued the spermatozoa into the female genital canal after coitus. Some researchers paid attention to spermiophagocytosis by leukocytes and macrophages in the uterine cavity (Thompson *et al.*, 1975), whereas others concentrated on possible morphological changes corresponding to the process of capacitation and acrosomal reactions (Motta and Blerkom, 1975). Sperm penetration into the ovum and the following processes of fertilization have been observed in rabbits, rats and some other mammals by SEM (Gould, 1975; Lin *et al.*, 1975; Sugawara *et al.*, 1975; Dudkiewicz and Williams, 1976).

SEM study of the *male reproductive organs* started only quite recently but this field seems to provide us with useful information which has been overlooked or difficult to obtain by the TEM.

The *seminiferous tubule* of the testis is composed of germ cells which proliferate and metamorphose into spermatozoa, and Sertoli cells which "cradle" the germ cells. A large number of spermatozoa deriving from the same spermatogonium become mature attaining the cytoplasmic bridges which connect them. These intercellular bridges, which cause a synchronical differentiation of germ cells, are impressively demonstrated under the SEM even in spermatids which have completed maturation (Plate VI-2B; Gravis, 1978). The processes and mechanisms as to how the Sertoli cell "cradles" the germ cells, anchors the spermatids while incorporating their residual bodies, and at last releases the matured spermatozoa into the tubular lumen have been extensively studied by TEM and seem worth attacking by SEM also (Gravis, 1978; Johnson, 1978).

The knowledge of the structure of the *interstitial tissue of the testis* has been much advanced by the SEM studies of Connell (1976) and Clark (1976). Especially the beautiful work of the latter author in the rat testis revealed the peculiar three-dimensional arrangement of the interstitial tissue and lymphatic sinusoid, though the TEM study by Fawcett *et al.* (1973) indicates that the interstitial lymphatic space is extensively developed only in rodents.

The three-dimensional extension and epithelial features of the human *rete testis* were explored through a combined SEM and TEM study by Roosen-Runge and Holstein (1978). The *ductuli efferentes, ductus epididymidis* and *ductus deferens* were studied by SEM in the dog and bull (Nowell and Faulkin, 1974) and in the rat (Hamilton *et al.*, 1977) with special regard to the regional variation in the epithelial cell types and surface fine structure. The structure of the stereocilia covering the epithelium of the epididymal duct has been impressively demonstrated by SEM (Murakami *et al.*, 1975; Hamilton *et al.*, 1977). As to the SEM observation of the ductus deferens, reports by Brueschke *et al.* (1974) and by Berns *et al.* (1974) are also available.

The seminal vesicle in the monkey was examined by SEM with special reference to the spermatozoa immigrating into the organ and phagocytosed by intraluminal macrophages (Murakami *et al.*, 1977, 1978). The SEM views of the prostate and urethra of the dog were reported by Nowell and Faulkin (1974).

Plate VI-1
A. Crosswise Fractured Seminiferous Tubule.   Rat.

The *germinal epithelium* of a seminiferous tubule is flatly cracked. The occurrence of mature spermatids extending their long tails into the lumen, together with the oval, immature spermatids (St) and rounded germ cells underneath, indicates that this tubule segment is at stage VIII. The whorled arrangement of sperm tails in the tubule is characteristic of rodents.
   × 900

B. Fractured Germinal Epithelium.   Rat.

A closer view of the rat tubule at a similar stage is demonstrated. At the lower left mature spermatids are seen with their long heads and thick tails. Spermatids of immature stage (St) are seen in the middle, connected by intercellular bridges. The thin, smooth fibers labeled asterisks probably correspond to the juvenile tails of the immature spermatids. The Sertoli cells show complicated profiles because they are manifoldly ramified and lamellated. The areas containing white balls (arrows) are interpreted as residual bodies incorporated into the Sertoli cells. Lipid droplets in the bodies deeply osmificated apparently emit more secondary emission than other parts of the tissue.
   × 3,300

(Plate VI-1: Courtesy of Dr. K. Goto, Departments of Urology and Anatomy, Niigata University School of Medicine)

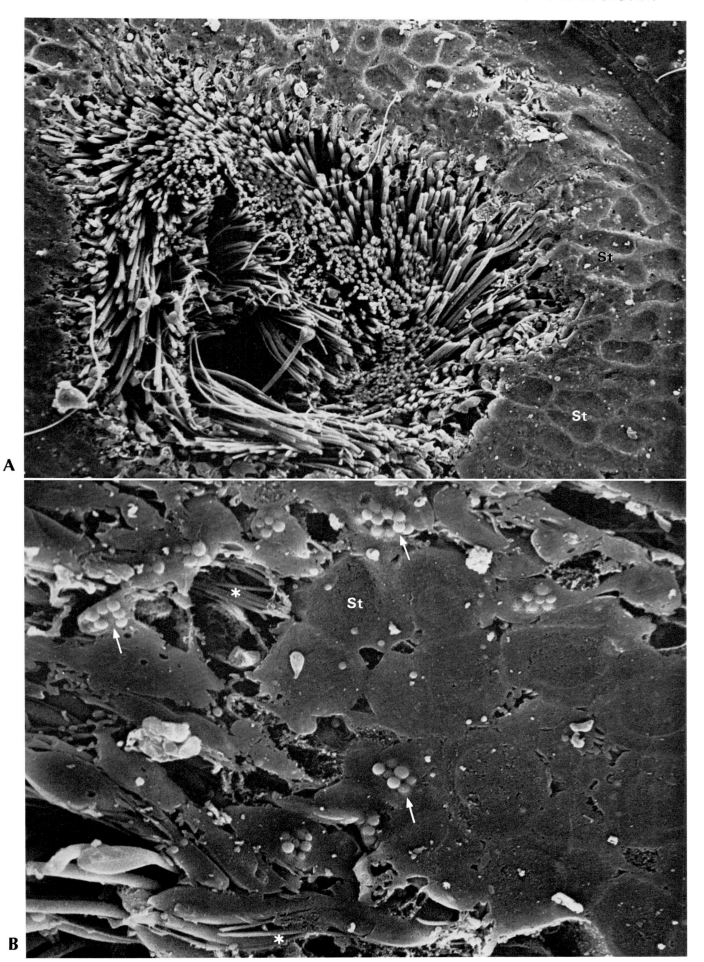

Plate VI-2   Maturing Spermatids.   Rat.

Different portions of maturing spermatids in the rat seminiferous tubule of probable stage VIII are demonstrated after freeze cracking.

A shows *spermatids* connected to lamellated cytoplasm of *Sertoli cells* (S). The main portion of the sperm tail is smooth and cylindrical but the proximal portion is thickened, covered by a cytoplasmic droplet. This cytoplasm is extended by a slender bridge (arrows) which connects with a residual body of the spermatid (R). The residual bodies are characterized by numerous lipid droplets and are ensheathed by the cytoplasm of Sertoli cells.

The thin filaments labeled an asterisk are presumed to be juvenile tails from the more deeply located, immature spermatids.

B provides a closer view of maturing rat spermatids. The middle piece (M) shows a periodically undulated surface which corresponds to the mitochondria winding around this portion of the tail.

The spermatids in this micrograph are connected by an intercellular bridge (arrow).

C reveals the crosswise fracture of spermatid tails. The construction of the tail is well known from TEM studies. An axial filament complex is surrounded by outer longitudinal fibers which, in cross section, appear like petals. These are further surrounded by ring fibers whose side aspect is viewed in part of the tails after the surface cytoplasm was peeled off by fracture.

The thin filaments apparently containing only the central filament complex are juvenile tails from immature spermatids.

A: X 4,800,   B: X 6,000,   C: X 11,000

(Plate VI-2: Courtesy of Dr. K. Goto, Departments of Urology and Anatomy, Niigata University School of Medicine)

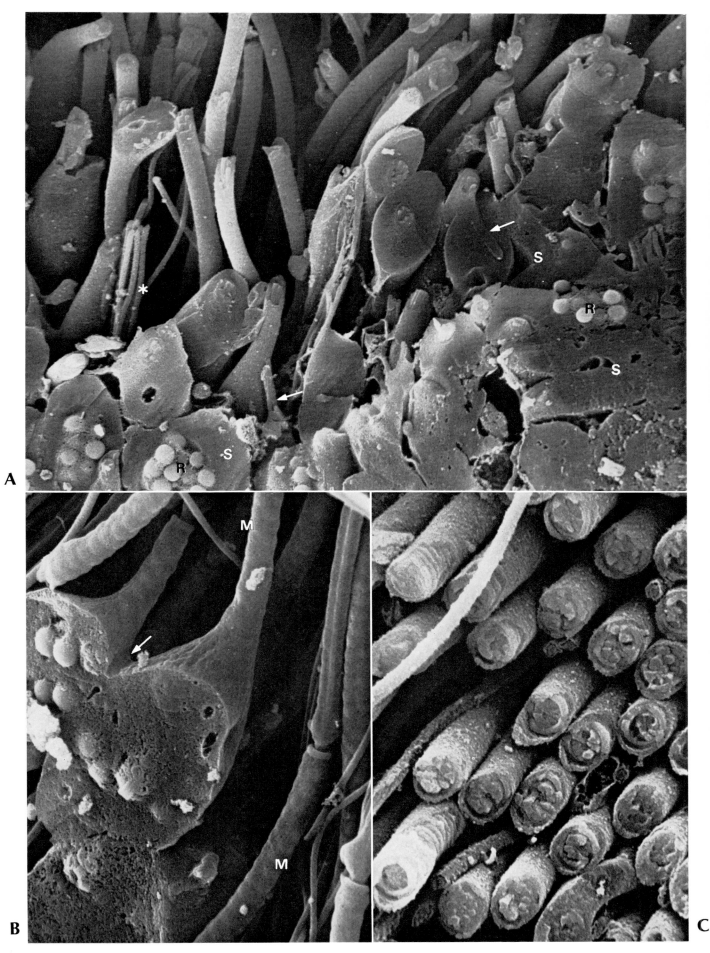

Plate VI-3
A. Maturing Spermatids. Human.

Spermatids are attached to the germ epithelial surface, apparently closely before spermiation, i.e., before leaving the Sertoli cells (S). The flattened head of the sperm is followed by an oval cytoplasmic droplet (D). This obscures the outline of the middle piece (M) but the boundary between the latter and the main portion of the tail (the site of the annulus) is generally clear (arrows).

The seminiferous tubule lumen contains numerous balls which may represent, at least partly, the germ cells wasted after atretic degeneration.

The testis was obtained from a 70-year-old man in castration performed for treatment of prostatic cancer.

X 8,500

B. Ejaculated Spermatozoa. 27-Year-Old Man.

Although ejaculates of healthy fertile men usually contain a considerably high population (up to 25–30%) of atypical spermatozoa, this micrograph demonstrates one of the normal human spermatozoa.

The head is the nuclear portion of the spermatozoon and its apical smooth half indicates the portion equipped with the *acrosome* (A). This contains different lytic enzymes to be used when the spermatozoon penetrates through the cells and membranes surrounding the ovum. The funnel-shaped portion posterior to the acrosome contains the *postnuclear cap* beneath the relatively rough cell surface. The undulated *middle piece* (M) contains mitochondria, the energy source of the spermatozoon. The more smoothly cylindrical main portion of the tail follows the middle piece.

X 15,000

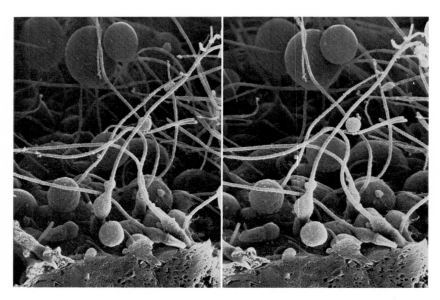

Fig. VI-1 Human Seminiferous Tubule in Stereo.
The stereo pair shows the same view as Plate VI-3A.
X 2,000

(Plate VI-3A and Fig. VI-1: From K. Goto: *Biomed. Res.* 2, Suppl. :361–374, 1981)
(Plate VI-3B: Courtesy of Dr. K. Goto, Departments of Urology and Anatomy, Niigata University School of Medicine)

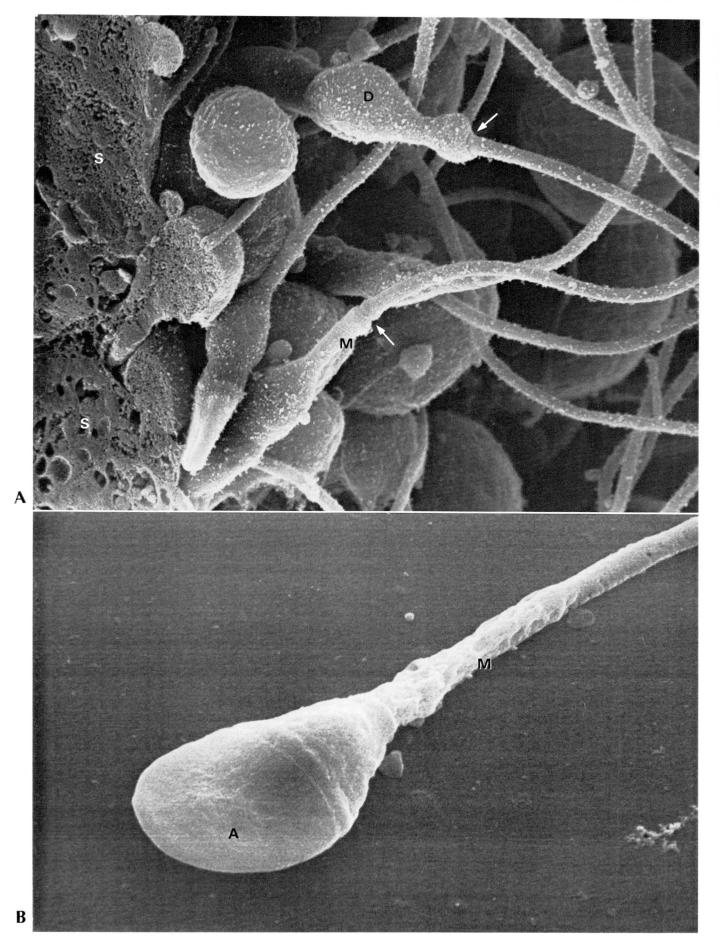

Plate VI-4   Abnormal Spermatozoa.   Ejaculates of Suspected Infertile Men.

A shows one of the most common abnormalities of spermatozoa with the *cytoplasmic droplet,* which should have been abandoned during the maturation of the spermatozoon. The boundary between the middle piece and the main part of the tail is distinct in this spermatozoon.

B shows a form whose *middle piece* is smooth and thinned in its distal half. This indicates that mitochondria are lacking in this part. The proximal half of the middle piece is thickened but this apparently is due to the remaining cytoplasmic droplet. The SEM view does not reveal whether some mitochondria occur in this part or not. Needless to say, absent or reduced mitochondria in the middle piece causes complete or incomplete *immotility of spermatozoa.*

C represents one of the typical deformities of spermatozoa which may be called a *golfclub* form. In this disturbance the flat head and the folded tail mostly lie on the same plane. The golfclub spermatozoa occur usually in patients of suspected sterility and seldom in fertile men.

D is a *double-headed* form with a confluent middle piece. The acrosomes and mitochondria seem to be well formed.

E is a *double-tailed* spermatozoon. Mitochondria spirally cover the bipartite middle piece.

F is a *microcephalic* type which usually characterizes severe male infertility. The head is vestigial and the middle piece mitochondria are completely absent.

G is a *macrocephalic* form also indicative of severe infertility. Different shapes of the head, sometimes really monstrous, are known.

H represents a spermatozoon with a spiral tail fused by a membrane. This deformity is not an artifactual product because the ejaculate of this infertile patient showed many spermatozoa of identical structure.

    A: × 7,800,　B: × 8,800,　C: × 10,000,　D: × 8,700

    E: × 4,600,　F: × 8,300,　G: × 8,300,　H: × 9,200

A,E and F:　Suspected infertile patient, 29 years old, sperm count: normal; atypical spermatozoa: 20%.
B, D and G:　Suspected infertile patient, 40 years old, sperm count: normal; atypical spermatozoa: 30%.
C: Suspected infertile patient, 31 years old, oligospermia (1800); atypical spermatozoa: 50% .
H: Suspected infertile patient, 26 years old, sperm count: normal, atypical spermatozoa: 50%.

(Plate VI-4: Courtesy of Dr. K. Goto, Departments of Urology and Anatomy, Niigata University School of Medicine).

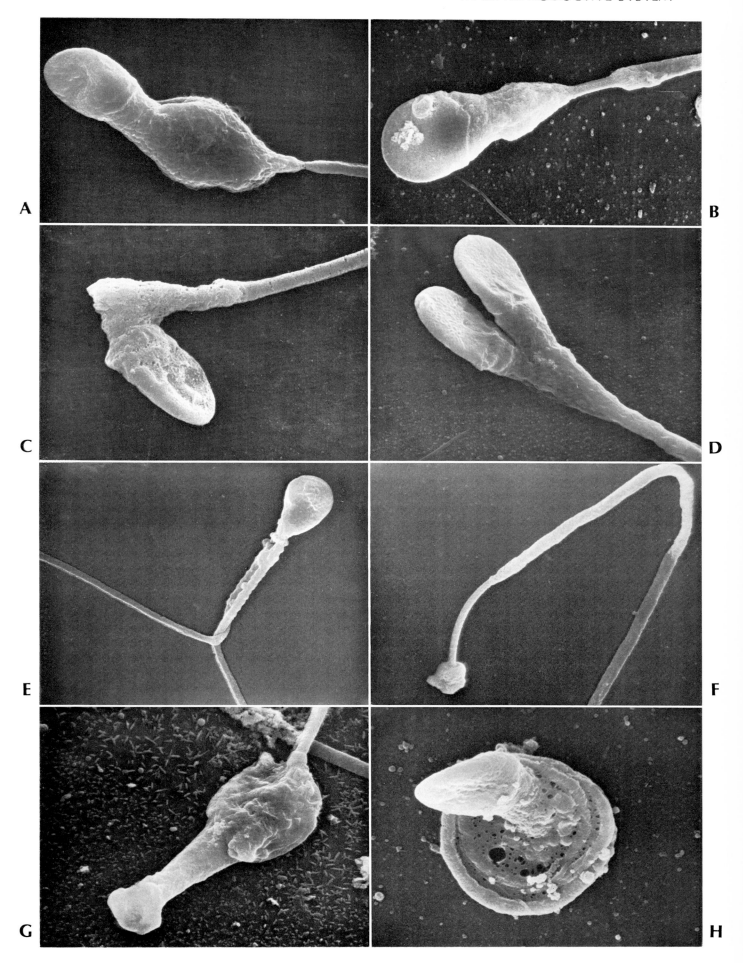

Plate VI-5   Rete testis.   Human.

The human *rete testis* is a network of irregular channels located between the seminiferous tubules of the testis and the ductuli efferentes of the epididymis. A recent study revealed that the rete is a complex myoelastic sponge (Roosen-Runge and Holstein, 1978) and the functions of its characteristically specialized portions are unknown.

A is an overview of the rete testis showing connected spaces of variable sizes. The spaces are traversed by ridges and bridges. The bridges, as shown in the middle of this micrograph, are mostly thin (5—30 μm in thickness) and columnar, and may be called *chordae retis* according to Roosen-Runge and Holstein (1978). The TEM observation of these authors indicates that the cordae contain a peculiar tissue consisting of central myoid cells and peripheral fibroelastic elements.

B shows a surface view of the channels of the rete. Squamous epithelial cells covered by irregular microvilli, as in this micrograph, line the major part of the rete including the surface of the chordae. Spermatozoa are seen on the epithelial surface.

C demonstrates another type of rete epithelium, which is represented by patches of prismatic cells of unknown function. The apical aspect of the cells shows a rather smooth swelling with a cilium (arrows) and marginally gathered, irregular microvilli.

A: X 220,   B: X 2,800,   C: X 2,400

(The specimens used for this plate were prepared by Dr. K. Goto of the Departments of Urology and Anatomy, Niigata University School of Medicine, from testes obtained in castration performed for treatment of prostatic cancer in a 79-year-old (A and B) and a 72-year-old (C) patient).

(Plate VI-5B and C: Reproduced from K. Goto: *Biomed. Res.* 2, Suppl. :361—374, 1981)

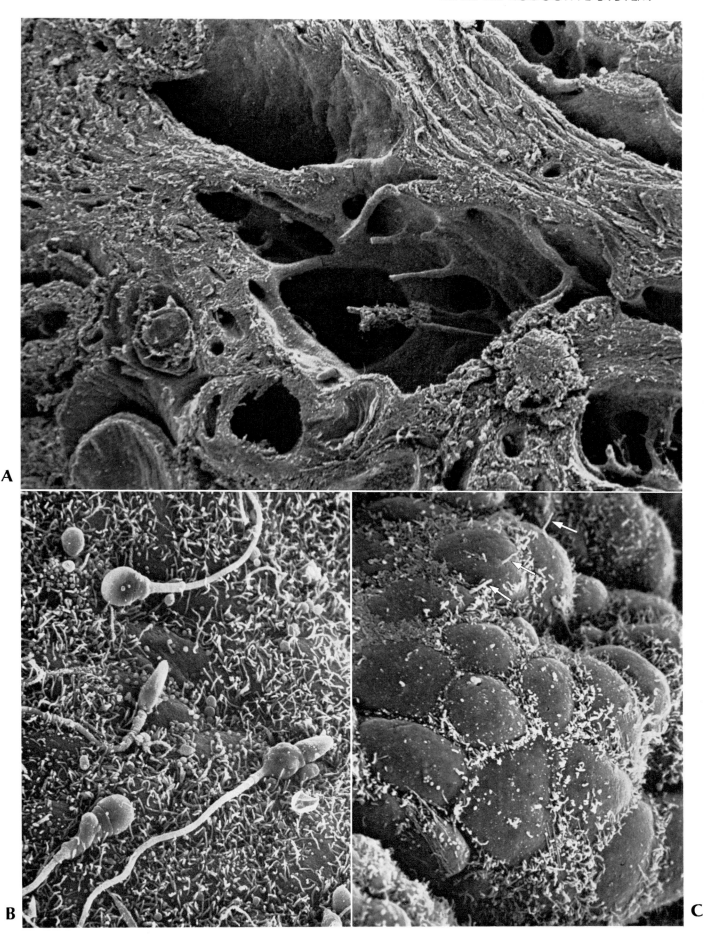

Plate VI-6 Ductuli efferentes. Human.

About 10 tightly wound tubules, *ductuli efferentes,* connect the rete testis with the ductus epididymidis. Secretory and reabsorptive functions of the ductule epithelium have been suggested. Two types of epithelial cells, i.e., *ciliated* and *non-ciliated* cells, generally are known but the SEM examination of human ductuli efferentes reveals the occurrence of two different non-ciliated elements.

A shows an overview of fractured ductuli efferentes. In the middle the characteristic pattern of epithelial cells in the efferent ductule is evident: Groups of high prismatic cells alternate with groups of quite low cells. Ciliated cells appear as white dots in this low-power micrograph.

B shows a part of the epithelial surface exhibiting ciliated cells and non-ciliated, microvillous cells. The short microvilli densely covering the latter cells show a conspicuous tendency to adhere to one another. These sticky microvilli have been noticed in the rat ductuli efferentes (Hamilton *et al.,* 1977).

C shows another cell population of the human ductule. Among the ciliated cells one may see domes of non-ciliated cells which are only thinly studded with villous and granular microprojections.

A: × 270,   B: × 2,200,   C: × 4,000

(Plate VI-6: Reproduced from K. Goto: *Biomed. Res.* **2**, Suppl.: 361–374, 1981)

Fig. VI-2 Cilia of Human Ductulus efferens in Stereo.
As seen also in Plate VI-6B and C, numerous small vesicles are attached to the cilia. The nature of these vesicles is unknown, though they may possibly represent secretory products of the epithelial cells.
 × 1,500

(The specimens used for this plate and figure were prepared by Dr. K. Goto, of the Departments of Urology and Anatomy, Niigata University School of Medicine, from the testes obtained in castration performed for treatment of prostatic cancer in a 65-year-old (A and C) and a 79-year-old (B and stereo) patient).

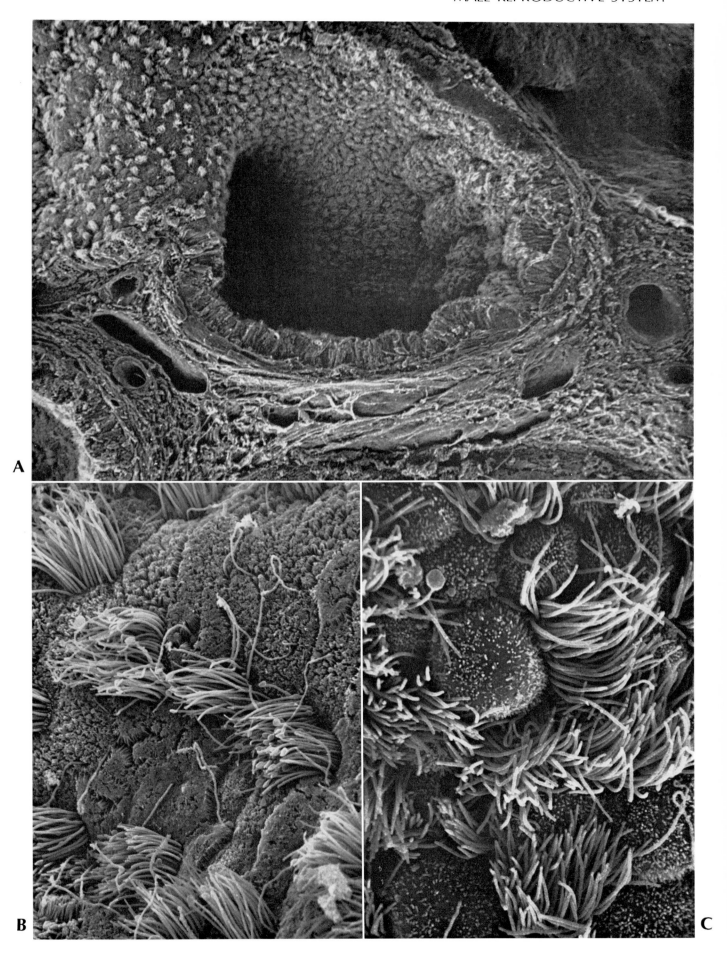

Plate VI-7

A. Ductus Epididymidis. Human.

The ductus epididymidis is a single but strongly convoluted duct. It receives semen from the ductuli efferentes and transfers it to the ductus deferens.

The epithelium (E) is composed of high prismatic cells which possess long microvilli called *stereocilia*. The apical cytoplasm is rich in vacuoles which presumably are involved in a secretory activity of the cells. Stereocilia are so called because they are long like cilia but do not beat actively like cilia. A less conspicuous movement of stereocilia, however, is suspected. A possible specific action of stereocilia upon spermatozoa remains also to be elucidated.

The ductus epididymidis, together with the ductus deferens, possesses a strong muscular wall (M) whose contraction causes ejaculation of the sperm stored in these ducts. The adventitia contains numerous blood vessels, especially veins (V).

X 500

B. Stereocilia. Japanese Monkey.

This micrograph shows how apical papillae of the epithelial cells of the epididymal duct extend tufts of stereocilia. A somewhat different structure of the roots of stereocilia seen in the human epididymal duct has been shown in Plate I-5B

X 4,300

(Plate VI-7A: The specimen was prepared by Dr. K. Goto of the Departments of Urology and Anatomy, Niigata University School of Medicine, from the testis obtained in castration performed for treatment of prostatic cancer in a 65-year-old patient).
(Plate VI-7B: Courtesy of Prof. M. Murakami and Dr. T. Shimada, Department of Anatomy, Kurume University School of Medicine.)

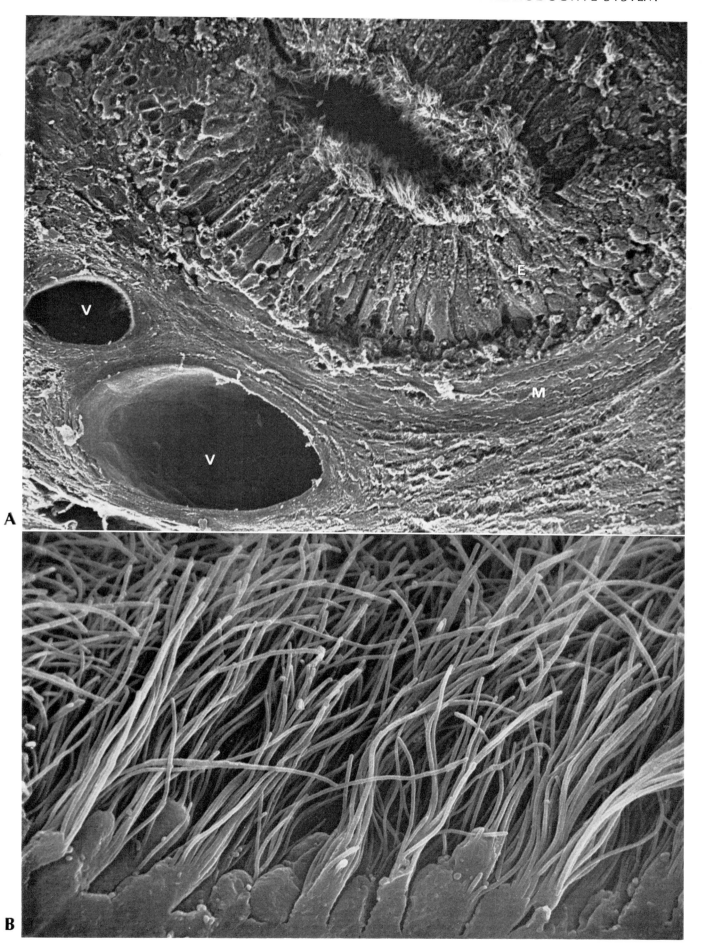

Plate VI-8   Seminal Vesicle.   Japanese Monkey.

The seminal vesicle is lined by a high prismatic epithelium which secretes a yellow, sticky fluid. This is delivered to the ejaculatory duct during ejaculation and serves as the vehicle and nutrients for the spermatozoa.

A shows the surface of the high prismatic epithelium lining the lumen of the seminal vesicle. High ridges of the mucous membrane form many deep grooves. Individual epithelial cells are identified as slight swellings in this low magnification. A few spermatozoa are attached, which apparently have migrated into the vesicle.

B is a closer view of the mucosal surface. The convex, hexagonal apices of the epithelial cells are covered by short microvilli. In some cells microvilli are attached to each other presumably by a sticky secretion of the cell. No cilia are found on the epithelial cells.

C shows a macrophage phagocytosing spermatozoa which have erroneously migrated into the vesicle. One may clearly see some heads and tails of the spermatozoa still left outside of the macrophage.

A: X 1,100,   B: X 6,600,   C: X 3,400

(Plate VI-8A: Courtesy of Prof. M. Murakami and Dr. Shimada, Department of Anatomy, Kurume University School of Medicine).
(Plate VI-8B and C: From M. Murakami, A. Sugita, T. Shimada and T. Yoshimura: *Arch. histol. jap.* **41**: 275—283, 1978).

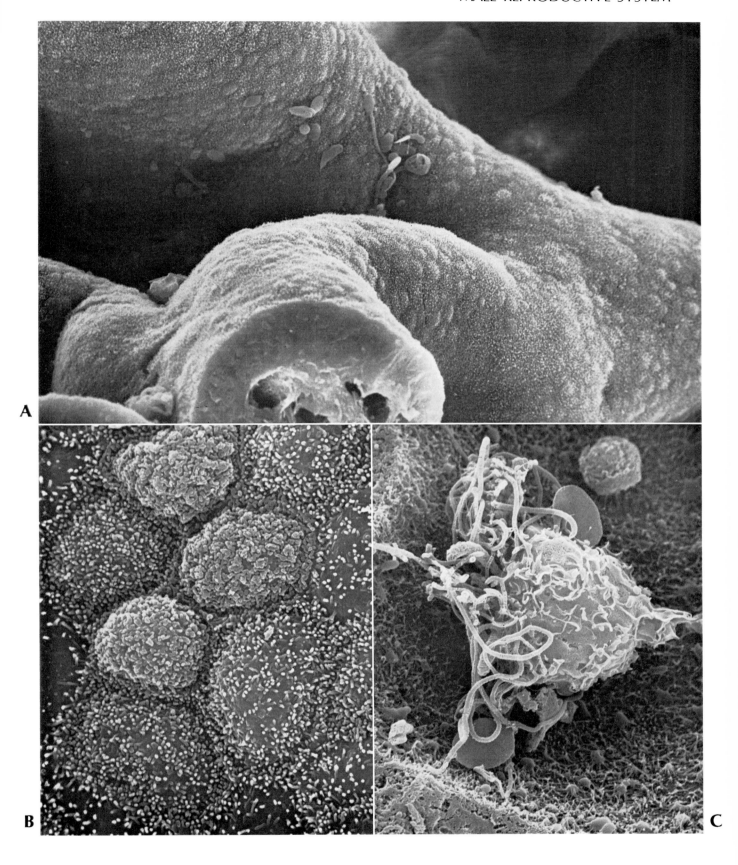

Plate VI-9   Penis and Urethra.   Japanese Monkey.

The composition of the penis is shown by a SEM view of its cross section (Fig. VI-3). The dorsal, major portion is the *corpus cavernosum penis* (c), the erection organ, surrounded by a thick collagen layer, *tunica albuginea* (a). The ventral, small portion represents the *corpus spongiosum* surrounded by a much weaker albuginea. The flattened *urethra* is indicated by an arrow.

A shows the luminal surface of one of the venous caves forming the corpus cavernosum penis. Flat endothelial cells with an oval nuclear swelling pave the surface. Microvilli tend to gather in the cell margins; the marginal folds are well developed.

B The *urethra* in the corpus penis is lined by a stratified columnar epithelium, which is fractured in this micrograph. The cell apices are irregularly polygonal and slightly convex.

C shows the apical aspect of the urethral epithelium in closer view. Short microvilli densely cover the cells.

A: × 3,100,   B: × 2,100,   C: × 6,800

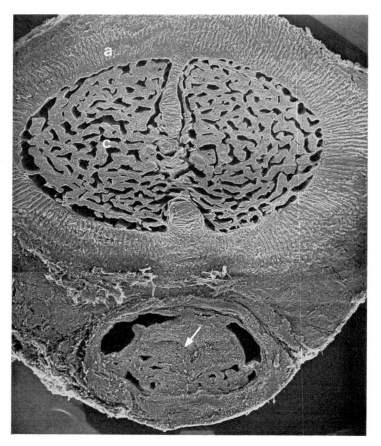

Fig. VI-3 Cross Section of Corpus Penis. Japanese Monkey.
For labels see the text above.
   × 8.8

(Plate VI-9 and Fig. VI-3: Courtesy of Prof. M. Murakami and Dr. T. Shimada, Department of Anatomy, Kurume University School of Medicine).

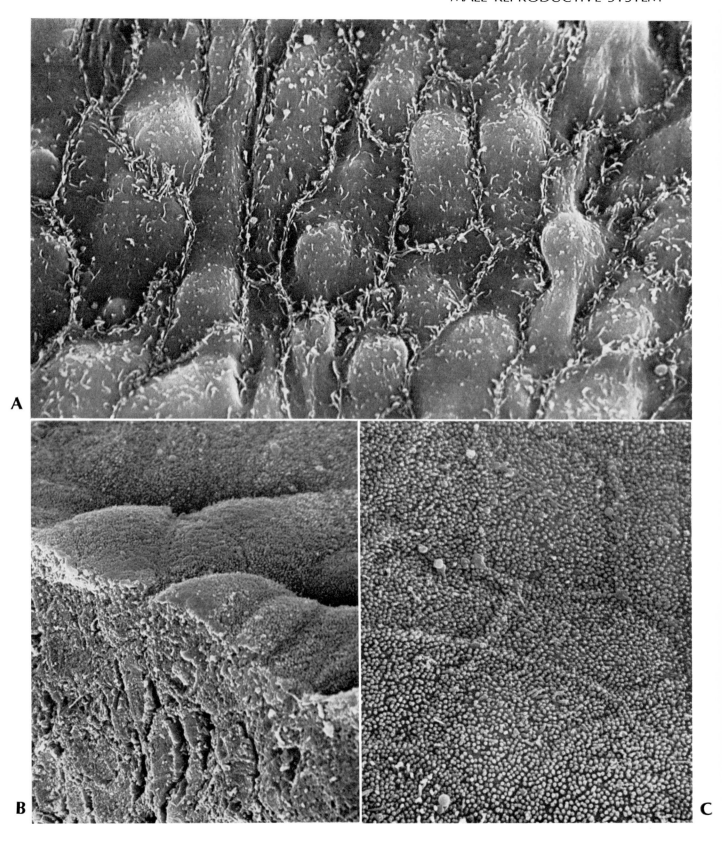

References

Berns, D. M., L. J. D. Zaneveld, R. A. Rodzen and E. E. Brueschke: Vasa deferentia of the human and dog: A study with the S.E.M. In: (ed. by) O. Johari and I. Corvin: Scanning Electron Microscopy/1974. IIT Research Institute, Chicago, 1974 (p.648—654).

Brueschke, E. E., J. J. D. Zaneveld, R. Rodzen and D. Berns: Development of a reversible vas deferens occlusive device. III. Morphology of the human and dog vas deferens: A study with the scanning electron microscope. *Fertil. Steril.* **25**: 687—702 (1974).

Clark, R. V.: Three-dimensional organization of testicular interstitial tissue and lymphatic space in the rat. *Anat. Rec.* **184**: 203—225 (1976).

Connell, C. J.: A scanning electron microscope study of the interstitial tissue of the canine testis. *Anat. Rec.* **185**: 389—402 (1976).

Dott, H. M.: Preliminary examination of bull, ram and rabbit spermatozoa with the stereoscan electron microscope. *J. Reprod. Fertil.* **18**: 133—134 (1969).

Dudkiewicz, A. B. and W. L. Williams: Interaction of rabbit sperm and egg. Examination of the same cryofractured specimens with scanning and transmission electron microscopy. *Cell Tiss. Res.* **169**: 277—287 (1976).

Fawcett, D. W., W. B. Naeves and M. N. Flores: Comparative observations on intertubular lymphatics and the organization of the mammalian testis. *Biol. Reprod.* **9**: 500—532 (1973).

Fujita, T.: Abnormal spermatozoa and infertility (man). In: (ed. by) E. S. E. Hafez: Scanning Electron Microscopic Atlas of Mammalian Reproduction. Igaku Shoin, Tokyo, 1975 (p.82—87).

Fujita, T., M. Miyoshi and J. Tokunaga: Scanning and transmission electron microscopy of human ejaculate spermatozoa with special reference to their abnormal forms. *Z. Zellforsch.* **105**: 483—497 (1970).

Goto, K.: The surface morphology of the epithelium of the human seminiferous tubule, rete testis, ductuli efferentes and ductus epididymidis. *Biomed. Res.* 2, Suppl. : 361—374 (1981).

Gould, K. G.: Mammalian fertilization. In: (ed. by) E. S. E. Hafez: Scanning Electron Microscopic Atlas of Mammalian Reproduction. Igaku Shoin, Tokyo, 1975 (p.288—299).

Gravis, C. J.: A scanning electron microscopic study of the Sertoli cell and spermiation in the Syrian hamster. *Amer. J. Anat.* **151**: 21—38 (1978).

Hamilton, D. W., G. E. Olson and T. G. Cooper: Regional variation in the surface morphology of the epithelium of the rat ductuli efferentes, ductus epididymidis and vas deferens. *Anat. Rec.* **188**: 13—28 (1977).

Johnson, L.: Scanning electron and light microscopy of the equine seminiferous tubule. *Fertil. Steril.* **29**: 208—215 (1978).

Lacy, D., A. J. Pettitt, J. M. Pettitt and B. S. Martin: Application of scanning electron microscopy to semen analysis of the sub-fertile man utilising data obtained by transmission electron microscopy as an aid to interpretation. *Micron* **5**: 135—173 (1974).

Lin, T. P., R. H. Glass, R. Bronson, J. Florence and M. Maglio: Interspecies sperm-egg interaction. In: (ed. by) E. S. E. Hafez: Scanning Electron Microscopic Atlas of Mammalian Reproduction. Igaku Shoin, Tokyo, 1975 (p.300—305).

Lung, B. and G. F. Bahr: Scanning electron microscopy of critical point dried human spermatozoa. *J. Reprod. Fertil.* **31**: 317—318 (1972).

Matano, Y., K. Matsubayashi, A. Ōmichi and K. Ohtomo: Scanning electron microscopy of mammalian spermatozoa. *Gunma Symp. Endocrinol.* **13**: 27—48 (1976).

Motta, P. and J. Van Blerkom: A scanning electron microscopic study of rabbit spermatozoa in the female reproductive tract following coitus. *Cell Tiss. Res.* **163**: 29—44 (1975).

Murakami, M., T. Shimada, C.-T. Huang and I. Obayashi: Scanning electron microscopy of epididymal ducts in the Japanese monkey (*Macacus fuscatus*) with special reference to the architectural analysis of stereocilia. *Arch. histol. jap.* **38**: 101—107 (1975).

Murakami, M., T. Shimada and K. Suefuji: Scanning electron microscopic observation of sperimophage cell within the lumen of the epididymal duct of the vasectomized Japanese monkey (*Macacus fuscatus*). *Experientia* **33**: 1101—1102 (1977).

Murakami, M., A. Sugita, T. Shimada and T. Yoshimura: Scanning electron microscope observation of the seminal vesicle in the Japanese monkey with special reference to intraluminal spermiophagy by macrophages. *Arch. histol. jap.* **41**: 275—283 (1978).

Nowell, J. A. and L. J. Faulkin: Internal topography of the male reproductive system. In: (ed. by) O. Johari and I. Corvin: Scanning Electron Microscopy/1974. IIT Research Institute, Chicago, 1974 (p.639—646).

Roosen-Runge, E. C. and A. F. Holstein: The human rete testis. *Cell Tiss. Res.* **189**: 409—433 (1978).

Sugawara, S., S. Takeuchi and E. S. E. Hafez: Sperm penetration. In: (ed. by) E. S. E. Hafez: Scanning Electron Microscopic Atlas of Mammalian Reproduction. Igaku Shoin, Tokyo, 1975 (p.280—287).

Thompson, J. E., R. J. Goodall, S. Sugawara and E. S. E. Hafez: Phagocytosis of spermatozoa. In: (ed. by) E. S. E. Hafez: Scanning Electron Microscopic Atlas of Mammalian Reproduction. Igaku Shoin, Tokyo, 1975 (p.88—98).

CHAPTER **VII**

FEMALE REPRODUCTIVE SYSTEM

The surface of the *ovary* is paved with polygonal mesothelial cells covered by microvilli and frequently provided with a cilium and occasionally blebs and ruffles. SEM studies by Motta (1974) in the rabbit suggest that these structures reflect their own cell cycle on one hand and the reproductive cycle on the other hand. The possibility of the mesothelial cells producing granulosa cells and interstitial cells of the ovary with their invaginations (crypts and cords) of proliferative capacity (Motta, 1974) may be effectively analysed by combined SEM and TEM studies.

The SEM observation of fractured structures of the ovary (Plates VII-1—4) has revealed interesting images of the granulosa cells with their feet extended into the zona pellucida, and the three-dimensional architecture of the corpus luteum (Van Blerkom and Motta, 1978) has been published. The process of ovulation in the rabbit has been extensively observed by SEM (Van Blerkom and Motta, 1979).

The *oviduct* is lined by a single prismatic epithelium represented by intermingled ciliated and secretory cells (Plate VII-6). Changes in their activities according to phase of the sexual cycle have provided one of the most favored objects of SEM observation (Ferenczy *et al.*, 1972; Patek *et al.*, 1972a, b; Rumery and Eddy, 1974).

The human uterotubal junction has been observed by SEM (Fadel *et al.*, 1976). This portion together with the isthmus of the oviduct in man and some animals represent critical barriers forcing adjustment of the travel of the impetuous spermatozoa so that they might arrive at the appropriate time at the site of fertilization, i.e., the ampulla of the oviduct (Bedford, 1970).

The *uterus* has been studied by SEM mainly with respect to the surface structure of the endometrium (Hafez, 1972; Kanagawa *et al.*, 1972; Anderson *et al.*, 1975; Motta and Andrews, 1976; Van Blerkom and Motta, 1979). In contrast to the oviduct the endometrial epithelium contains a more or less inconspicuous population of ciliated cells. The main cell type is secretory in nature and covered by irregular microvilli. The secretions of the endometrium, which is conspicuously enhanced in the progestational stage, are mainly produced by the winding and tubular invagination of the epithelium, the uterine glands.

As for the *vagina*, the interest of SEM researchers has concentrated on the rabbit vaginal epithelium as it is beautifully covered by microvilli and phagocytose numerous spermatozoa after coitus (Phillips and Mahler, 1977) (Plate VII-8).

In the female genital tract, the spermatozoa undergo special changes called *capacitation* so that they become able to fertilize the ovum. It has been disputed as to whether capacitation is reflected in any morphological changes of the spermatozoa such as in the surface fine structure. Motta and Van Blerkom (1975) correlated the eroded surface structure in the acrosome which they found in the intrauterine spermatozoa of the rabbit with a process in capacitation.

The process of fertilization as well as the cleavage, growth and differentiation of the ovum and embryo have provided marvelous objects for SEM studies. However, this atlas will omit all of them and will show some micrographs of the human *placental villi*, on which reference to some previous studies by SEM is also recommended (Okudaira *et al.*, 1972, Kaufmann *et al.*, 1979).

Plate VII-1
A. Primary Follicle. Rabbit Ovary.

Primary follicles are gathered in the cortical portion of the ovary. They consist of an *oocyte* with a large nucleus (N) and relatively homogeneous cytoplasm (C), and a flattened layer of follicle epithelial cells (E). Interstitial cells of irregular shape and collagen fibrils embed the follicles. A few blood capillaries (B) are opened.
  X 3,600

B. Secondary Follicle. Rabbit Ovary.

When the oocyte grows and its follicle epithelial cells become thickened, the follicle is called secondary. The follicle then continues to grow and the epithelial cells become multilayered and the membrane which surrounds the oocyte, called the zona pellucida (P), becomes thick and conspicuous.

This micrograph shows a later stage in the maturation of the secondary follicle. The *follicle epithelial cells* are piled up into multiple layers. The oocyte cytoplasm is now rich in organelles and lipid droplets, the latter causing the honeycomb appearance of the cytoplasm.
  X 600

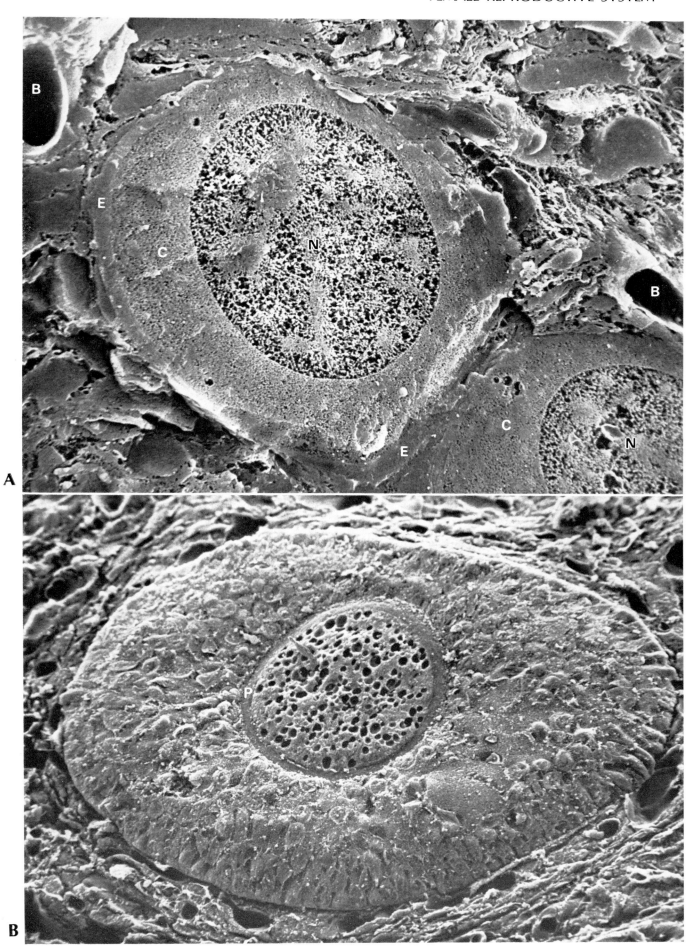

Plate VII-2   Follicle Epithelium and Theca.   Rabbit.

A peripheral part of a secondary follicle is shown. On the left hand the oocyte cytoplasm (C) is seen with numerous lipid droplets. The zona pellucida is destroyed in this preparation.

The *follicle epithelium* (E) consists of multilayered cells irregular in shape. The cells close to the oocyte are elongated radially. The intercellular space contains filamentous materials.

The layer outside is called the *theca* and characterized by circularly arranged collagen fibrils. The layer is divided into the *theca interna* (Ti) and *theca externa* (Te). The former contains relatively rich cells and blood capillaries (B) which are responsible for the production and transport of estrogen. In the theca externa, the collagen forms thicker fibers.

   X 1,500

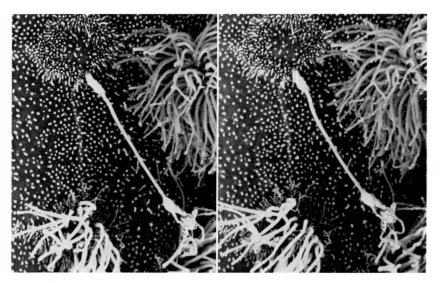

Fig. VII-1 Epithelium of Rete Ovarii in Stereo.   Rabbit.
In the medulla of the ovarium there is found a rudimental structure called the rete ovarii which is represented partly by epithelial cords and partly by irregular spaces. Epithelial cells lining one of those spaces are shown here. Possibly reflecting the rudimentary nature of the rete, one may find a *single cilium* peculiar in structure. A part of the epithelium is ciliated.

   X 3,300

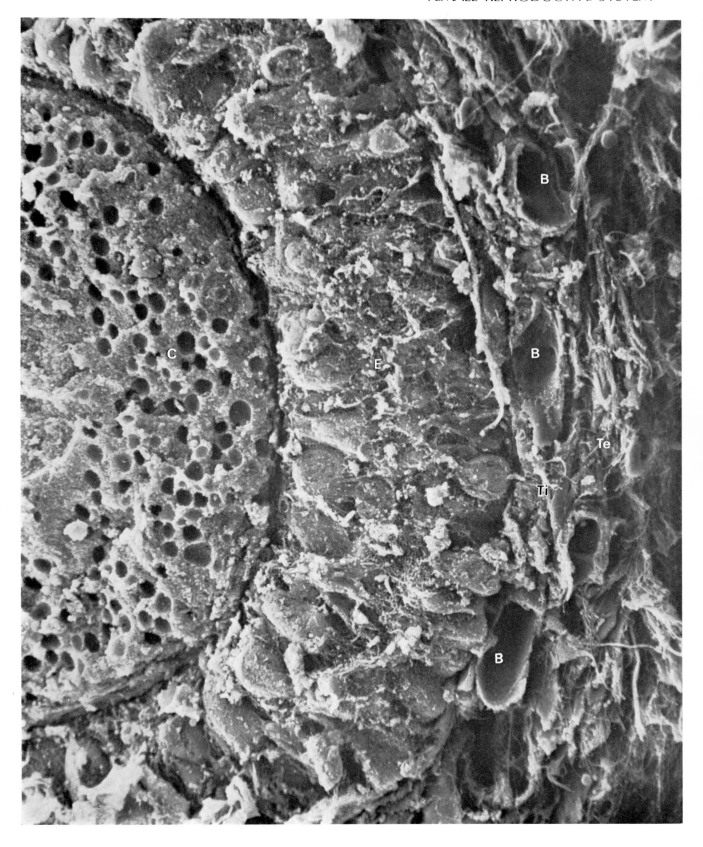

Plate VII-3   Graafian Follicle.   Mouse(A) and Cat(B).

When the secondary follicle grows larger and a space containing follicular fluid becomes obvious, it is called the Graafian follicle.

A demonstrates a multiple layer of follicle epithelial cells, now called the *stratum granulosum*, enclosing a large, crescent follicular space. The swollen portion of the stratum granulosum is the *cumulus oophorus* (C) and contains the *ovum* (primary oocyte), whose surface is exposed by fracture. The arrows indicate its capsule, the *zona pellucida* which is attached by the tapered processes of the follicle epithelial cells in one or two rows closest to the ovum. These rows of cells directly attached to the zona pellucida are called *corona radiata.*

B shows the surface view of the *zona pellucida* in a Graafian follicle. On the left hand the ovum is exposed. At the top and bottom the corona radiata cells remain inserting their feet into the pores of the zona pellucida (arrows). The surface of the zona appears labyrinthine by a network of ridges. Larger and smaller pores in its meshes receive the foot processes of the corona cells. A closer view is available in Plate VII-4C.

A: X 920,   B: X 16,000

(Plate VII-3: Courtesy of Prof. F. Yasuzumi, Department of Anatomy, Medical College of Okinawa).

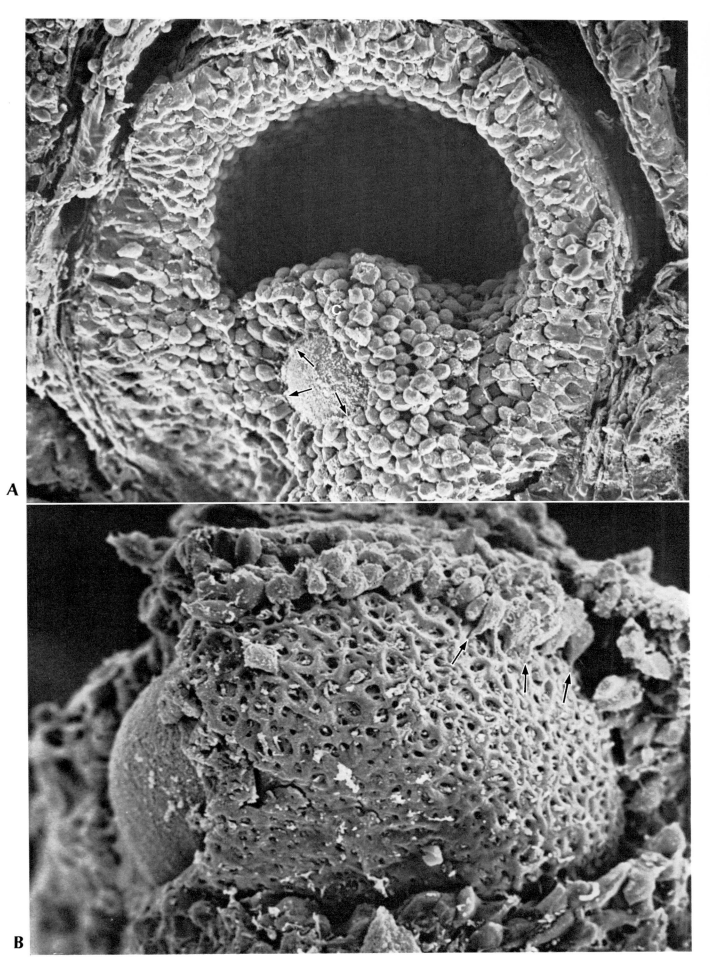

Plate VII-4   Stratum granulosum and Corona radiata.   Cat Graafian Follicle.

The follicle epithelial cells forming the stratum granulosum are loosely packed, polyhedral or ovoid cells which are believed to transport nourishment from the capillary-rich theca interna to the ovum. The innermost, corona radiata cells insert their foot-like processes into the zona pellucida and approach the microprocesses of the ovum within the pores of the zona.

A shows the cells in the *intermediate layer* of the stratum granulosum. The SEM reveals occurrence of thicker and thinner threads in the intercellular spaces. As more evident in the next figure, they are *cytoplasmic processes* of the follicle epithelial cells themselves. Their processes are entangled with each other and suggest an unknown intercellular substance transport or information exchange.

B demonstrates the *outermost layer* of the follicle epithelial cells forming the stratum granulosum. The cells touch the *basement membrane* (B) of the follicle with a stout process which is expanded like a web on the surface of the membrane. Behind the basement membrane, a small portion of the theca interna is shown. Note the delicate cytoplasmic processes issued from the ridges of the follicle epithelial cells (arrows).

C is a portion of the *zona pellucida*. The corona radiata cells have been removed, leaving some debris of their foot processes in the pores of the zona.

A: $\times$ 8,600,   B: $\times$ 10,000,   C: $\times$ 9,200

(Plate VII-4: Courtesy of Dr. F. Yasuzumi, Department of Anatomy, Medical College of Okinawa)

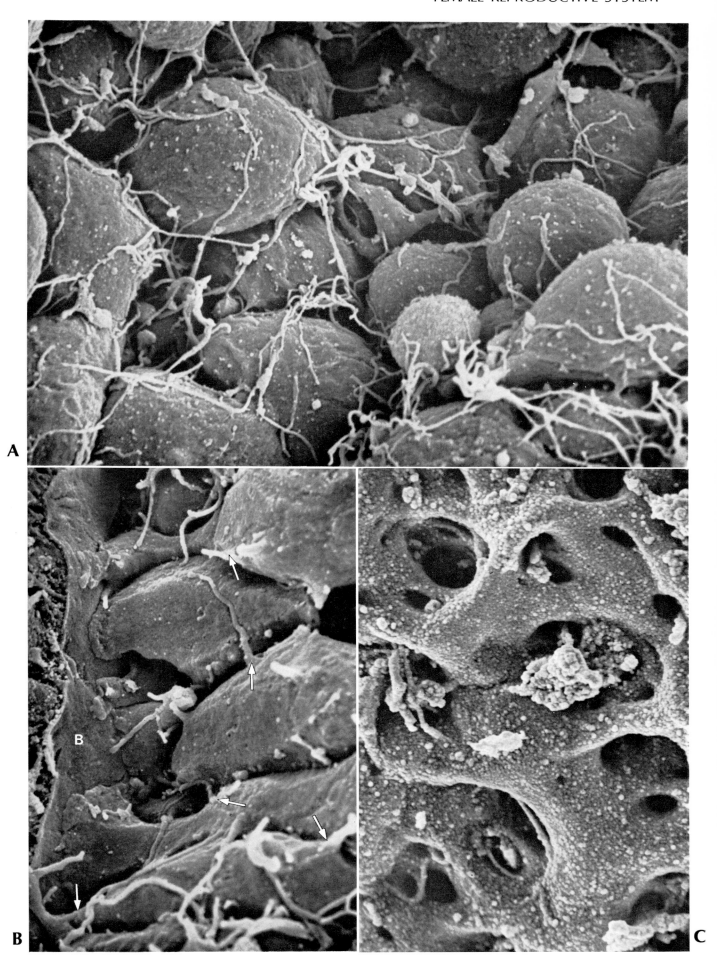

Plate VII-5   Oviduct.   Rabbit.

This is a low magnification view of the oviduct fractured crosswise. The thick epithelium is strongly folded into the lumen. Its luminal surface reveals round apical heads of secretory cells dispersed in the ciliated cells.

The *lamina propria* (Lp) is a reticular tissue rich in blood vessels, especially capillaries. The muscular layer of the oviduct is out of focus on the right.

The ovum is transported towards the uterus by means of the beat of the cilia and by the contraction of the musculature. If one estimates the diameter of rabbit ovum as 150 μm, it will correspond to a sphere 6 cm wide in this scanning micrograph.

X 400

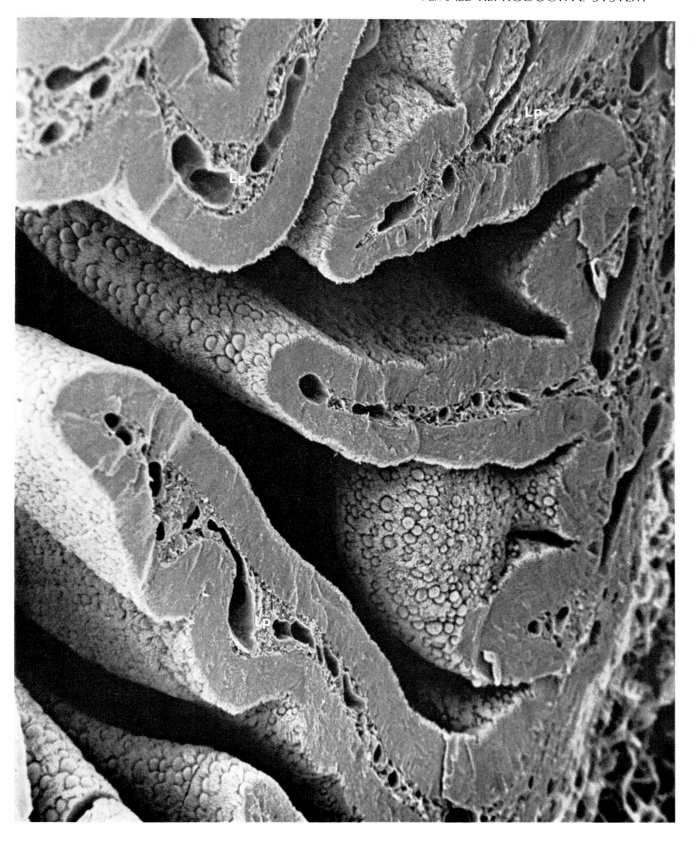

Plate VII-6 Epithelial Cells of Oviduct. Rabbit.

The micrograph shows the luminal surface of the oviduct in a sexually matured, virgin rabbit.

Two types of epithelial cells are clear. One is represented by *ciliated cells* with uniform cilia whose beating movement is directed from the ovarian end towards the uterus. The other type comprises *secretory cells* whose rounded apical portions covered by short microvilli protrude into the lumen in various sizes. The grade of the protrusion of the secretory cells as well as their secretory activity undergo a regular change during the sexual cycle.

The product of the secretory cells includes glycoprotein which is known to be indispensable for the maintainance and cleavage of the ovum. The secretion comes out of the cell by the swelling and breaking of portions of the apical surface.

X 4,000

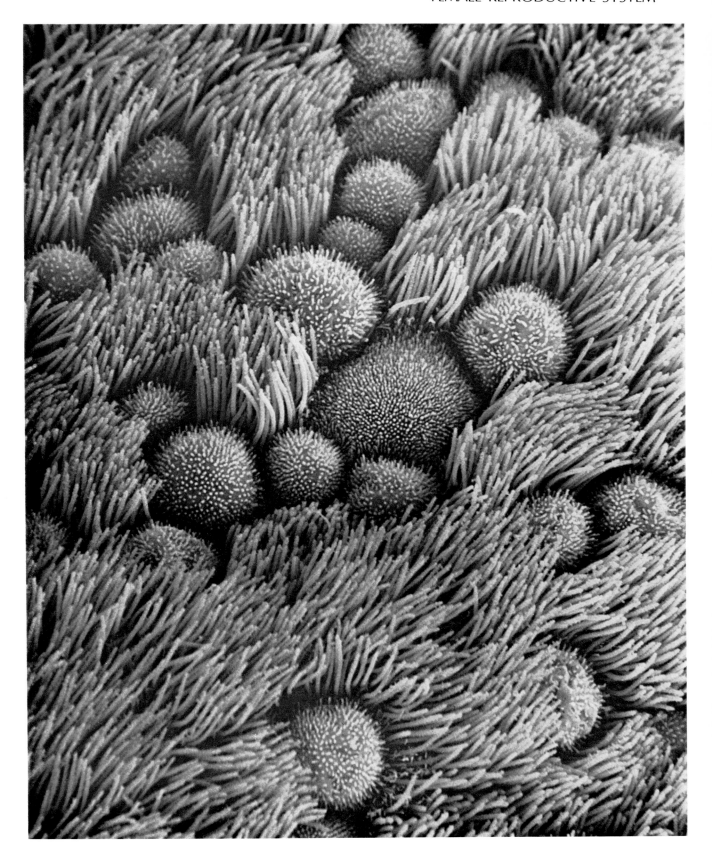

Plate VII-7   Uterus and Uterine Gland.   Rabbit.

The uterus in the rabbit, as in the human, is lined by a mucous membrane called the endometrium which mainly consists of a high prismatic epithelium and its deep tubular invaginations, the uterine glands.

This micrograph shows the orifice of a uterine gland and its surrounding paved with the epithelial cells. The cells are covered by fairly uniform microvilli. Some cells are covered by a mucous substance (*). The arrow indicates a small apocrine-like projection which might possibly be involved in the secretory activity of the cell.

Not only the uterine gland cells but also the endometrial epithelial cells are known to secrete mucous substances. Though not shown in this micrograph, a few ciliated cells are known, both in the rabbit and human, to occur in the endometrium.

Six hours after coitus.

X 6,000

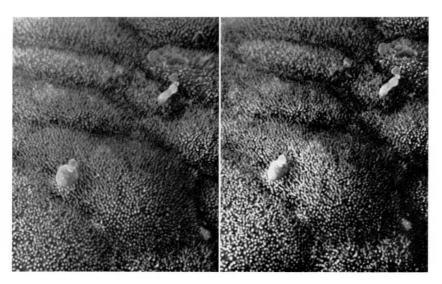

Fig. VII-2 Endometrial Epithelial Cells in Stereo.   Rabbits.

The apocrine-like projections mentioned above are demonstrated. These structures are presumed to represent a process in the release of the mucous secretions from the cells.

Six hours after coitus.

X 3,300

Plate VII-8   Vaginal Epithelium in Spermatophagocytosis.   Rabbit.

In human and most laboratory animals the vagina is lined by a stratified squamous epithelium, but in the rabbit the lining is a single prismatic epithelium.

This micrograph reveals both surface and intercellular aspects of the rabbit vaginal epithelium. The concave apical surface of the high prismatic cells is covered by microvilli, whereas their side surface is provided with plate-like processes for intercellular gearing (arrows). Moreover, some cells have been obliquely fractured in their apical cytoplasm, which reveals numerous secretory vesicles(∗).

Conspicuous of the rabbit vagina is the vigorous spermatophagocytosis by the epithelial cells. Two sperm heads (S) are shown being phagocytosed.

Six hours after coitus.

X 6,000

Fig. VII-3 Vaginal Spermatozoa in Stereo.   Rabbit.
Spermatozoa in the vagina are attached to the epithelial surface by their heads. The micrograph further demonstrates apocrine-like projections of the epithelial cells.

Six hours after coitus.

X 2,000

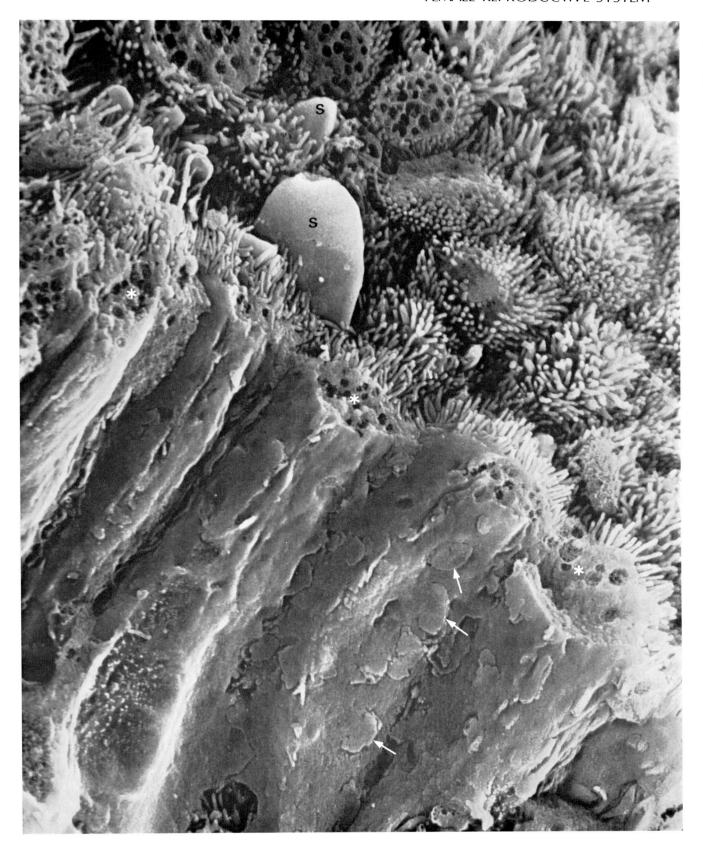

Plate VII-9 Placenta: Chorionic Villi. Human, 10 Months of Pregnancy.

The fetal life is supported by the vast area for substance exchange formed by the chorionic villi of the placenta. Here oxygen and nutritive substances are taken up from the maternal to the fetal circulation, whereas carbon dioxide and waste products of fetal metabolism are transferred from the fetal to the maternal blood.

A is an overview of chorionic villi. The maternal blood which had filled the intervillous spaces was washed away in specimen preparation.

B is a closer view of the velvet-like surface of the villus. Finger-shaped microvilli cover the surface and they enormously enlarge the substance exchange area.

A: × 640,   B: × 74,000

Fig. VII-4 Chorionic Microvilli in Stereo. Human, 2.5 Months of Pregnancy.
Chorionic villi may be covered by complex tufts of microvilli. It is noticed also that the microvilli shown here are swollen in their tips.
× 5,400

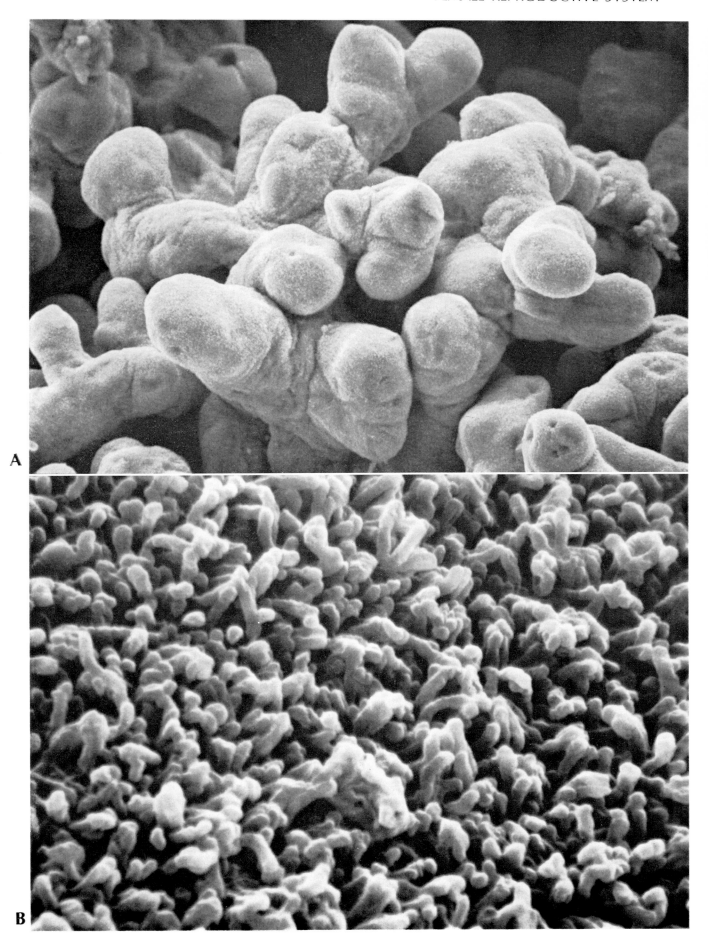

Plate VII-10   Placental Barrier.   Human, 3.5 Months of Pregnancy.

The surface epithelium of the placental villi is composed of a syncytiotrophoblast associated, from the basal side, with an incontinuous layer of cytotrophoblast (Langhans cells).

The apical structure of the syncytial layer is highly variable, showing different grades of enlargement of the surface area.

A demonstrates the strongly folded surface of the syncytial layer, each cytoplasmic ridge being covered by microvilli.

B shows the syncytial trophoblast (S) and a few Langhans cells (L). On the left hand is a blood capillary lined by a relatively thick endothelium and containing a fetal erythrocyte. The bar indicates the plate called the *placental barrier,* through which the gas and substance exchange occurs between the fetal and maternal blood.

The stroma of the villus is supported by reticular fibers whose large meshes contain fibroblast-like cells (F) and a kind of macrophage (Hofbauer cell), the latter not being demonstrated in this micrograph.

A: × 8,500,   B: × 3,500

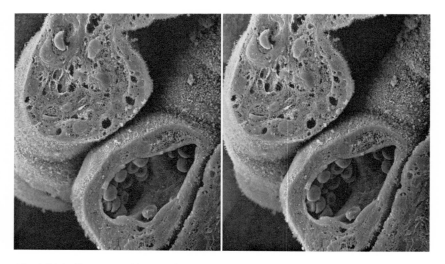

Fig. VII-5 Fractured Placental Villi in Stereo.   Human, 9 Months of Pregnancy.
The villus below shows a large blood capillary surrounded by a thinned and extended placental barrier. Numerous fetal erythrocytes are seen. The villus above includes some blood capillaries but they are not very clear as they contain coagulated plasma. A few erythrocytes are embedded there.

× 600

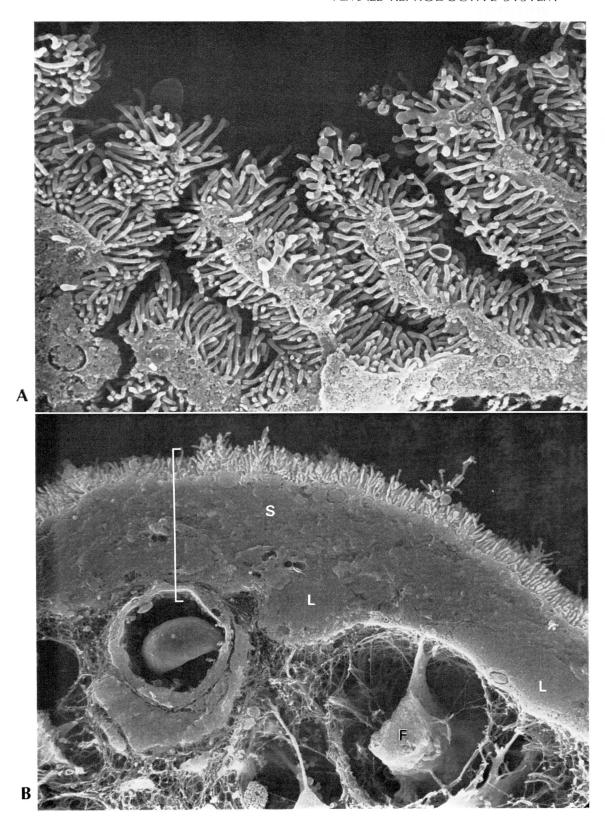

References

Anderson, W. A., Y.-H. Kang and E. R. DeSombre: Estrogen and antagonist-induced changes in endometrial topography of immature and cycling rats. *J. Cell Biol.* **64**: 692–703 (1975).

Bedford, J. M.: The saga of mammalian sperm from ejaculation to syngamy. In: (ed. by) H. Gibian and E. J. Plotz: Mammalian Reproduction. Springer-Verlag, Berlin· Heidelberg. New York, 1970 (p. 124–182).

Fadel, H. E., D. Berns, L. J. D. Zaneveld, G. D. Wilbanks and E. E. Brueschke: The human uterotubal junction: A scanning electron microscope study during different phases of the menstrual cycle. *Fertil. Steril.* **27**: 1176–1186 (1976).

Ferenczy, A., R. M. Richart, F. J. Agate, Jr., M. L. Purkerson and E. W. Dempsey: Scanning electron microscopy of the human fallopian tube. *Science* **175**: 783–784 (1972).

Hafez, E. S. E.: Scanning electron microscopy of rabbit and monkey female reproductive tract epithelium. *J. Reprod. Fertil.* **30**: 293–296 (1972).

Kanagawa, H., E. S. E. Hafez, W. C. Pitchford, C. A. Baechler and M. I. Barnhart: Surface patterns in the reproductive tracts of the rabbit observed by scanning electron microscopy. *Anat. Rec.* **174**: 205–226 (1972).

Kaufmann, P., D. K. Sen and G. Schweikhart: Classification of human placental villi. I. Histology. *Cell Tiss. Res.* **200**: 409–423 (1979).

Motta, P.: The fine structure of ovarian cortical crypts and cords in mature rabbits. A transmission and scanning electron microscopic study. *Acta anat.* **90**: 36–64 (1974).

Motta, P. and J. Van Blerkom: A scanning electron microscopic study of rabbit spermatozoa in the female reproductive tract following coitus. *Cell Tiss. Res.* **163**: 29–44 (1975).

Motta, P. M. and P. M. Andrews: Scanning electron microscopy of the endometrium during the secretory phase. *J. Anat.* **122**: 315–322 (1976).

Okudaira, Y., K. Hayakawa, N. Hamanaka, G. Ueda, S. Yoshinare, Y. Sato and K. Kurachi: Human placental villi: Scanning electron-microscopic observations. *Acta Obst. et Gynaec. Jap.* **19**: 109–117 (1972).

Patek, E., L. Nilsson and E. Johannisson: Scanning electron microscopic study of the human fallopian tube. Report I. The proliferative and secretory stages. *Fertil. Steril.* **23**: 459–465 (1972a).

Patek, E., L. Nilsson and E. Johannisson: Scanning electron microscopic study of the human fallopian tube. Report II. Fetal life, reproductive life, and postmenopause. *Fertil. Steril.* **23**: 719–733 (1972b).

Phillips, D. M. and S. Mahler: Phagocytosis of spermatozoa by the rabbit vagina. *Anat. Rec.* **189**: 61–72 (1977).

Rumery, R. E. and E. M. Eddy: Scanning electron microscopy of the fimbriae and ampullae of rabbit oviducts. *Anat. Rec.* **178**: 83–102 (1974).

Van Blerkom, J. and P. Motta: A scanning electron microscopic study of the luteo-follicular complex. III. Repair of ovulated follicle and the formation of the corpus luteum. *Cell Tiss. Res.* **189**: 131–154 (1978).

Van Blerkom, J. and P. Motta: The Cellular Basis of Mammalian Reproduction. Urban & Schwarzenberg, Baltimore, Munich, 1979.

CHAPTER **VIII**

SKIN

The skin consists of the *epidermis* which is a cornified, stratified epithelium, *dermis* which is a dense connective tissue, and *subcutaneous tissue* which is generally a loose connective tissue. Specialized structures of the epidermis form different kinds of accessory organs of the skin, like hairs, nails and sebaceous and sweat glands. The contents of this chapter will be restricted to a few micrographs of *epidermis and its appendages*, as the large variety of skin structures observed by SEM must form an independent atlas.

The *epidermis* is a pile of dead cell sheets and, during drying of the specimens, the cells are easily lifted and scaled. Crio-SEM, i.e. the technique of observing specimens kept frozen within the SEM is suited for obtaining low to middle power images closer to the natural stage of skin surfaces (Fujita and Tokunaga, 1975) (Fig. VIII-1).

The human epidermis observed in histological specimens is highly variable in structure in different parts of the body. If one observes the cells forming the superficial layer of epidermis by SEM, one will be again surprised to learn how variable they are in shape and surface structure according to the different portions of the body. An extensive and systematic work performed by Wolf (1939) using light microscopic techniques will be a good guide for SEM studies of epidermal cells. *Replicas* of skin surface and *epidermal cells peeled off* with a piece of cellophane tape or any other adhesive sheet may provide useful material for SEM observation; these methods are applicable to living subjects (Fujita *et al.*, 1969; Johnson *et al.*, 1970).

The surface fine structures of *hairs* have been extensively studied and the natural history, pathology and forensic medicine of the hair are being markedly enriched by application of SEM (Dawber and Comaish, 1970; Fujita *et al.*, 1971; Wyatt and Riggott, 1977). The nail surface is flat and uninteresting. *Sweat glands*, especially their orifices which vary in structure according to the body area, are interesting objects of SEM (Fujita, 1973). Beautiful SEM observations of the secretory activity of ceruminous apocrine glands (Kurosumi and Kawabata, 1976) and mammary glands (Murakami *et al.*, 1977) have been published.

Plate VIII-1   Hairs.   Forehead of a 50-Year-Old Male.

A shows a portion of a *lanugo*. This category of thin hair generally is accompanied by a large rosette of epidermal cells. The *rosette* is formed by cornified cell scales located at the edge of the invagination of the external root sheath.

The shaft of the lanugo is covered by hair cuticles of relatively large size.

B shows the *cuticles* of a thicker hair. The wavy edge of the cuticles is broken by the wear and tear of the hair surface.

A: X 2,500, B: X 7,800

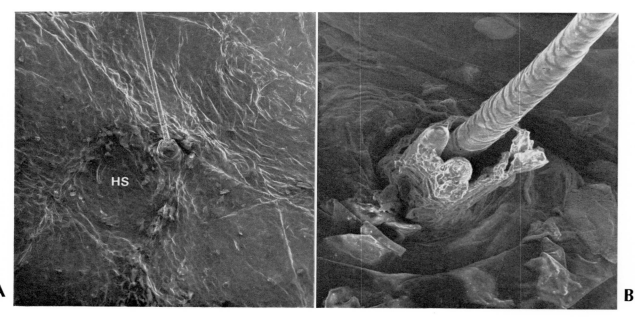

A   **B**

Fig. VIII-1 Crio-SEM Images of Human Skin.

These micrographs were obtained by scanning formalin fixed and frozen specimens. The structures of skin surface are preserved closer to natural state as the specimens have not been dried.

A. Skin from the lateral side of the shoulder. A thin hair (lanugo) is accompanied by a round plateau, *Haarscheide* (hair disc) of Pinkus (HS), which represents a special sensory apparatus for touch and pressure.

B. A completed lanugo and two young lanugos growing out of the hair follicle. From the extension side of the elbow.

A: X 80,   B: X 800

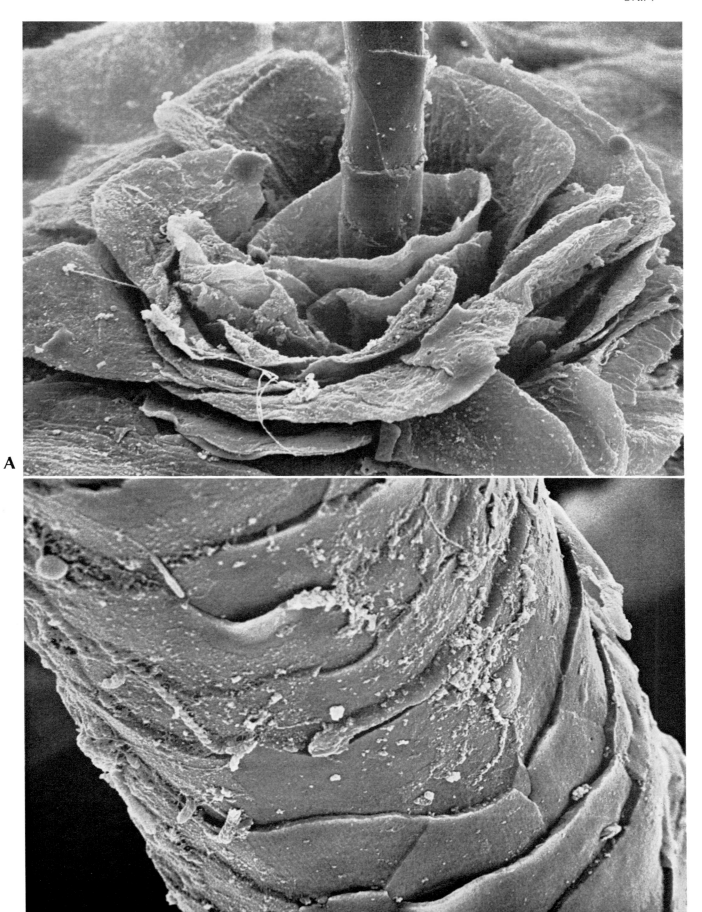

A

B

Plate VIII-2 Sweat Gland. Sole of a 12-Year-Old Girl.

The surface view of sweat pores differs conspicuously according to the portions of the body. In the palm they appear as spiral holes in a flat plate, whereas in the sole they are hidden in a concentric heap of epidermal cells.

A shows two skin ridges of the sole. A row of *sweat pores* (arrows) occurs in the middle of each ridge. The sweat pore is characterized by rose-like concentration of scaly cells.

B shows the portion of a sweat gland duct which spirally penetrates the epidermis. The duct was opened with a razor blade and the continuous, wrinkled wall was exposed.

A: × 200,   B: × 1,900

Fig. VIII-2 Sweat Pore in Stereo.   Sole of a 50-Year-Old Male.
The sweat duct in the epidermis of the sole is lined with a thin layer of specialized epidermal cells.
× 270

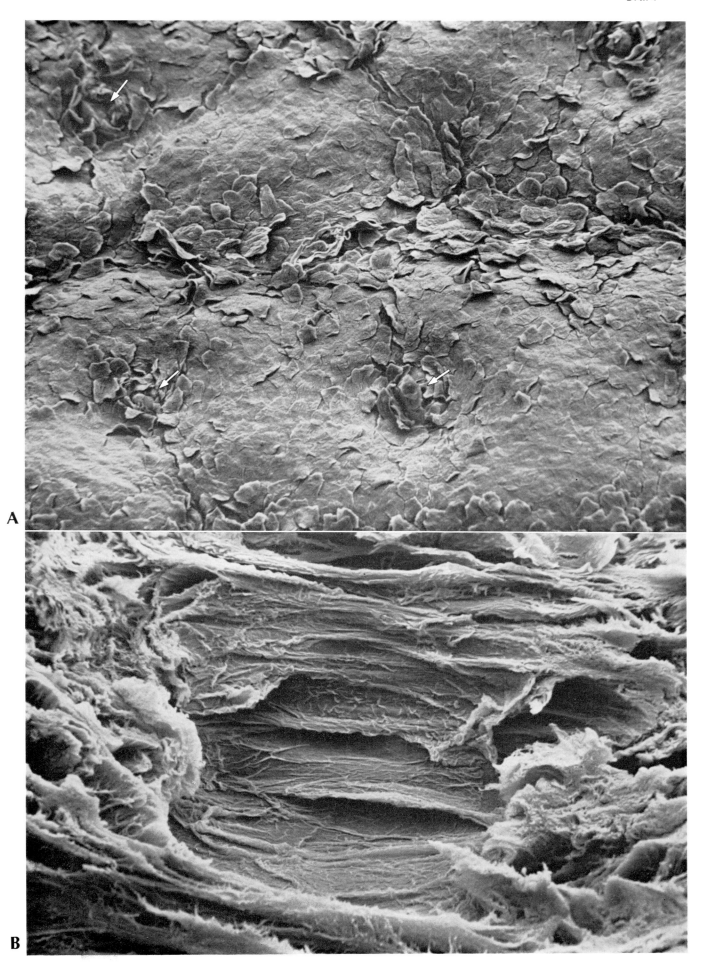

Plate VIII-3   Epidermis.   Finger of a 50-Year-Old Male.

This micrograph shows the backside (deeper side) of a scaled layer of the *stratum corneum* of the finger belly. The flattened cells of polygonal outline still show triangular and quadrangular facets reminiscent of the original polyhydral shape of the cells.

    The backside of the cells is covered by scale-like projections which are named *microfoliae* in this atlas (page 13; Plate I-8). The microfoliae are geared with holes on the superficial side of the adjacent cell.

    X 3,000

Plate VIII-4 Epidermis. Sole of a 50-Year-Old Male.

The *microfoliae* of the stratum corneum cells are best developed in the epidermis of the sole.

A shows long, tongue-like microfoliae on the deeper side of epidermal cells. The microfoliae tend to be fused into long ridges near the margin of the cells.

B demonstrates reticular *micro-ridges* on the opposite (superficial) side of the stratum corneum cells. The elongate holes formed between the ridges receive the microfoliae of the adjacent cell.

The sole epidermis prevents sliding between the flattened cells by this elaborate cell gearing.

A: X 4,700,   B: X 3,700

References

Dawber, R. and S. Comaish: Scanning electron microscopy of normal and abnormal hair shafts. *Arch. Derm.* **101**: 316—322 (1970).

Fujita, T.: Surface structure of skin. (In Japanese) In: (ed. by) M. Seiji, K. Kurosumi and Y. Mishima: Basic Dermatology. Asakura-Shoten, Tokyo, 1973 (p. 1—34).

Fujita, T. and J. Tokunaga: Scanning electron microscopy in histology and cytology. In: (ed. by) E. Yamada: Recent Progress in Electron Microscopy of Cells and Tissues. Igaku Shoin, Tokyo, 1975 (p. 319—342).

Fujita, T., J. Tokunaga and H. Inoue: Scanning electron microscopy of the skin using celluloid impressions. *Arch. histol. jap.* **30**: 321—326 (1969).

Fujita, T., J. Tokunaga and H. Inoue: Atlas of Scanning Electron Microscopy in Medicine. Igaku-Shoin, Tokyo, 1971. (p. 50—63).

Johnson, C., R. Dawber and S. Shuster: Surface appearances of the eccrine sweat duct by scanning electron microscopy. *Brit. J. Derm.* **83**: 655—660 (1970).

Kurosumi, K. and I. Kawabata: Transmission and scanning electron microscopy of the human ceruminous apocrine gland. I. Secretory glandular cells. *Arch. histol. jap.* **39**: 207—229 (1976).

Murakami, M., T. Shimada, T. Nishida and M. Sakima: Scanning electron microscopic study of the mammary gland of rats during and after lactation. *Arch. histol. jap.* **40**: 421—429 (1977).

Wolf, J.: Die innere Struktur der Zellen des Stratum desquamans der menschlichen Epidermis. *Z. mikrosk.-anat. Forsch.* **46**: 170—202 (1939).

Wyatt, E. H. and J. M. Riggott: Scanning electron microscopy of hair. Observations on surface morphology with respect to site, sex and age in man. *Brit. J. Derm.* **96**: 627—633 (1977).

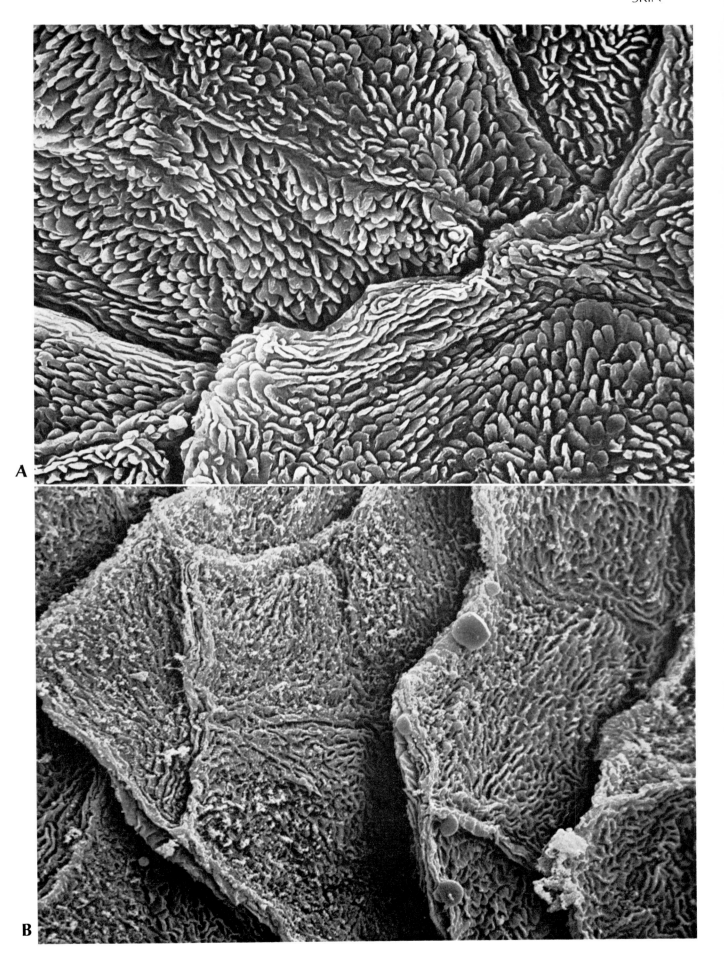

SENSORY ORGANS

Organ of Corti in the Inner Ear

The organ of Corti is a device of sound perception located on the *basilar membrane* of the cochlea. When sound is transmitted from the middle ear to the inner ear, vibration of the fluid (*perilymph*) in the cochlea occurs and this causes a delicate, vertical movement of the basilar membrane, which then is detected by the sensory cells by means of their dislocation against a gelatinous structure, the *tectorial membrane* which hangs over the cells.

As in the parts of the inner ear specialized for static and kinetic senses, the auditory cells in Corti's organ are *hair cells* exposed to the fluid called *endolymph,* and therefore have proved to be beautiful objects of SEM observation beginning with the pioneering papers published by Lim and Lane (1969) and Lim (1970). The hair cells form a complicated sensory epithelium together with a system of supporting cells (*pillar cells, phalangeal cells* of Deiters, etc.) and terminal fibers of the *cochlear nerve* which are characteristically naked in the spaces called the *inner tunnel* and *Nuel's space.* The three-dimensional extensions and relations of these structures, including the nerve terminals on the hair cells, have been demonstrated by appropriate fracture of Corti's organ (Bredberg, 1977; Shinozaki and Miyoshi, 1980).

The structure of the *tectorial membrane* has been studied by SEM (Tanaka *et al,* 1973; Hoshino, 1974; Ross, 1974), and special attention has been paid to the mode of connection or contact of the sensory hairs of the hair cells with the tectorial membrane, which is important for the understanding of the mechanism of audition (Hoshino, 1974, 1977).

SEM also revealed unequivocally the characteristic and selective *damage of the hair cells* after intense auditory stimulation (Lindeman and Bredberg, 1972; Lim, 1976; Hunter-Duvar, 1977) and after administration of ototoxic antibiotics (Tanaka *et al.,* 1973; Theopold, 1977; Hunter-Duvar, 1978; Leake-Jones and Vivion, 1979).

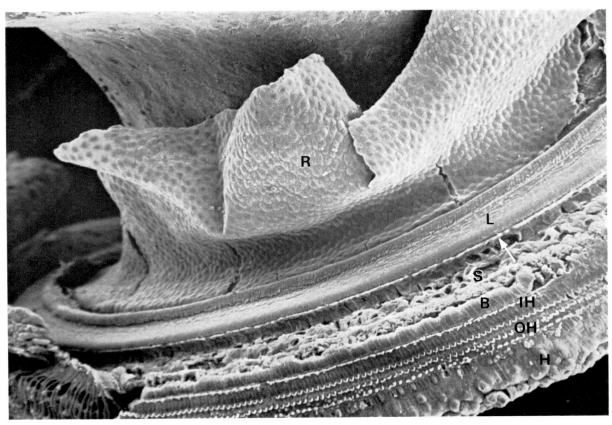

Fig. IX-1 Approach to the Organ of Corti. Guinea Pig.
The lower turn of the cochlear duct is opened and its floor, the organ of Corti, is exposed after removal of the tectorial membrane. Cushion-like swellings of *Hensen's cells* (H) are, interiorly, followed by three rows of *outer hair cells* (OH) and one row of *inner hair cells* (IH), then *border cells* (B) and *inner sulcus cells* (S). The edge labeled by the arrow is called *labium limbi vestibulare* and represents the site where the interior border of the tectorial membrane was attached. The flat plate, *limbus laminae spiralis* (L), forms the most interior or axial part of the cochlear duct floor. The roof of the cochlear duct is *Reissner's membrane* (R) which binds it against the scala vestibuli. In this specimen the membrane is torn and dislocated but the nuclear swellings of the squamous epithelial cells covering the membrane are clearly seen.

X 240

(Courtesy of Dr. Y. Igarashi, Department of Otorhinolaryngology, Niigata University School of Medicine).

Plate IX-1 Overview of Hair Cells. Guinea Pig.

One row of the *inner hair cells* (top) and three rows of the *outer hair cells* in an upper portion of the lower turn of the guinea pig cochlea are demonstrated. The sensory hairs of the inner hair cells which are arranged in a strongly flattened U shape are slightly disheveled in this preparation. The outer hair cells have their sensory hairs arranged in a V shape. This V is acuter in angle in the outer row cells and more obtuse in the inner row cells.

The hair cells are housed in a *reticular membrane* formed by the apical plates of supporting cells: *phalangeal cells of Deiters* and *inner* and *outer pillar cells*. The arrangement of microvilli may facilitate the distinction of the cells.

X 4,400

(Plate IX-1: Courtesy of Dr. Y. Igarashi, Department of Otorhinolaryngology, Niigata University School of Medicine).

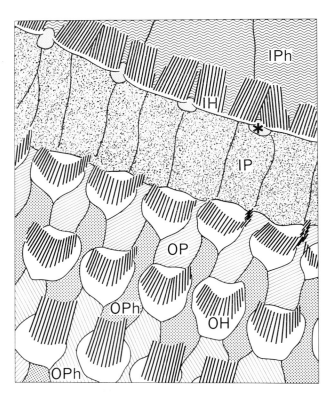

Fig. IX-2 Diagram of the Sensory and Supporting Cell Arrangement.

This diagram was made by tracing the SEM image on the opposite page to show the regular pattern of the *reticular* membrane of supporting cells housing the inner (IH) and outer hair cells (OH).

IPh: inner phalangeal cells and their phalanges (*), IP and OP: head plates of inner and outer pillar cells; OPh: phalanges of outer phalangeal cells.

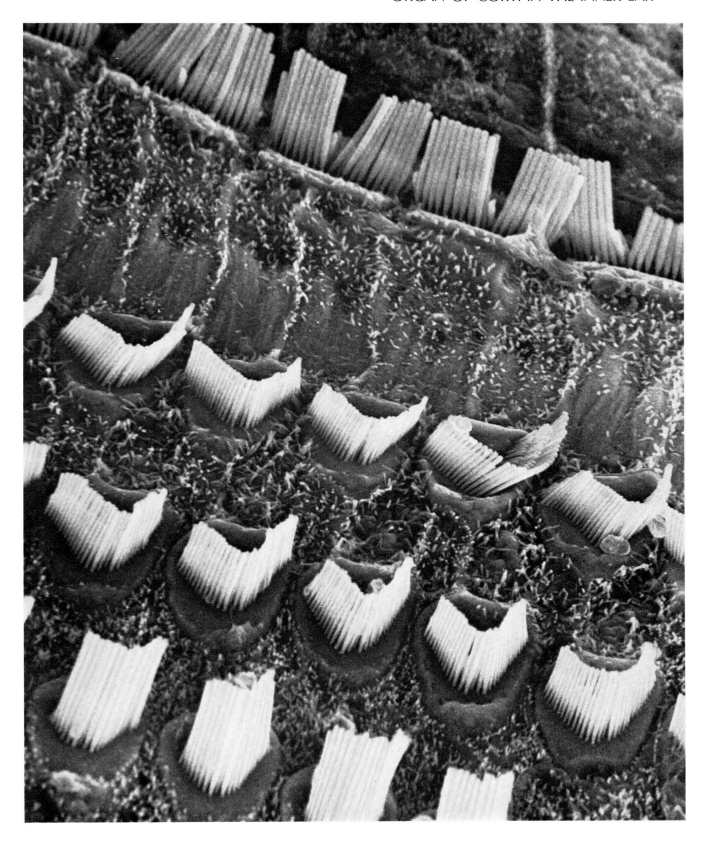

Plate IX-2   Closer View of Hair Cells.   Guinea Pig.

The micrographs demonstrate the outer hair cells of the first (interior) row in an upper portion of the lower cochlear turn.

A shows the smooth apical surface of the hair cells with V-arranged sensory hairs. These are *stereocilia,* i.e., microvilli of a specialized form. A real cilium or *kinocilium* occurs in the cochlear hair cells in the fetus but it degenerates after birth leaving a vestigeal *basal body*, whose swelling can be recognized in this micrograph (arrows).

B reveals the stereocilia in closer view. They are arranged in three rows. Behind the main or front row of long stereocilia, two further rows of gradually shorter and weaker stereocilia are seen. Note the thinned basal portion of all the hairs and the characteristically tapered heads of the middle row hairs.

A: X 11,000,   B: X 18,500

(Plate IX-2: Courtesy of Dr. Y. Igarashi, Department of Otorhinolaryngology, Niigata University School of Medicine).

Plate IX-3   Supporting Device of Hair Cells.   Rhesus Monkey.

The hair cells of Corti's organ in these micrographs reveal partly their apical aspect facing the tectorial membrane and partly their basal parts exposed by fracture.

A is an overview showing the outer hairs of the second and third rows, the first row being hidden under the *tectorial membrane* (T). Beneath the fractured cochleal membrane, the columnar cell bodies of the hair cells are screened by obliquely extended *phalangeal processes of Deiters cells* (D).

B shows a closer aspect of the preceding view. The base of the hair cell column is covered by the body of a *Deiters cell* (D) from which a long *phalangeal process* (P) projects to support, by a web-like expansion called the *phalanx*, the upper edge of a hair cell. The adjacent phalanges are connected with each other and form the reticular membrane (page 260).

The phalangeal processes obliquely cross the hair cell columns, as they are regularly inclined towards the apex of the cochlea. Thus, the phalangeal cell in the rhesus monkey skips two or three hair cells of the third row while it extends from its base to the phalanx (Hoshino, 1975). The grade of inclination of the phalangeal processes and the number of hair cells skipped thereby are known to differ considerably by species and also by the rows of the cells (Held, 1926).

A: × 1,650,   B: × 4,100

(Plate IX-3: Courtesy of Dr. T. Hoshino, Department of Otorhinolaryngology, Hamamatsu Medical College).

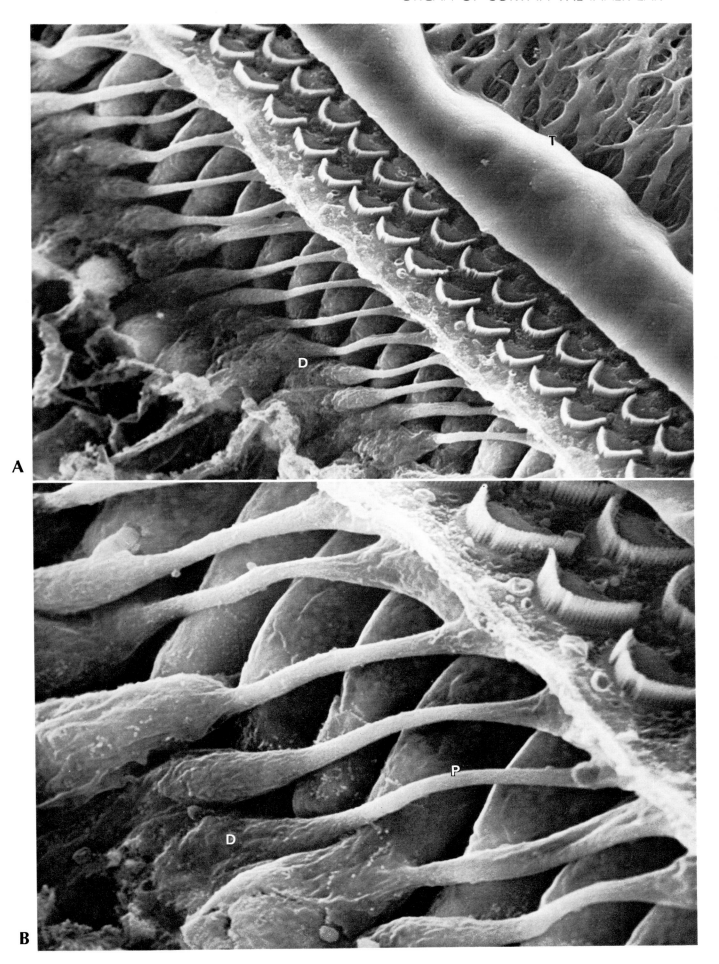

Plate IX-4
A. Nerve Terminals on Hair Cells.   Rabbit.

The columns shown here are *outer hair cells* of the second row in the basal (first) turn of the rabbit cochlea. The basal portions of the hair cells are covered by *Deiters' cells* (D); small finger-like processes may fringe the margin of the cells.

Spoon-like *nerve terminals* of a larger and a smaller size are seen in contact with the hair cells. According to Shinozaki and Miyoshi (1979) *the large-sized terminals* (Nl) are *efferent* in nature and their axons are characterized by beady swellings. *The small-sized terminals* (Ns) are believed to be *afferent* in nature; their axons are thin and lacking in swellings.

X 6,000

B. Reissner's Membrane.   Guinea Pig.

The very thin Reissner's membrane is composed of a flat *endothelium* facing the scala vestibuli and a single squamous *epithelium* facing the cochleal duct, a narrow connective tissue space being sandwiched in between. This micrograph is a flat view of the epithelium, showing the polygonal cell boundaries and short microvilli.

X 2,600

C. Tympanic Covering Layer of the Basilar Membrane.   Guinea Pig.

The basilar membrane is lined by a *mesothelium* which is characteristically loose in structure in the upper turn of the cochlea as shown in this micrograph. The cells are mostly bipolar, possessing long, thin processes.

X 2,500

(Plate IX-4A: Courtesy of Prof. M. Miyoshi and Dr. Naoko Shinozaki, Department of Anatomy, Fukuoka University School of Medicine).
(Plate IX-4B and C: Courtesy of Dr. Y. Igarashi, Department of Otorhinolaryngology, Niigata University School of Medicine).

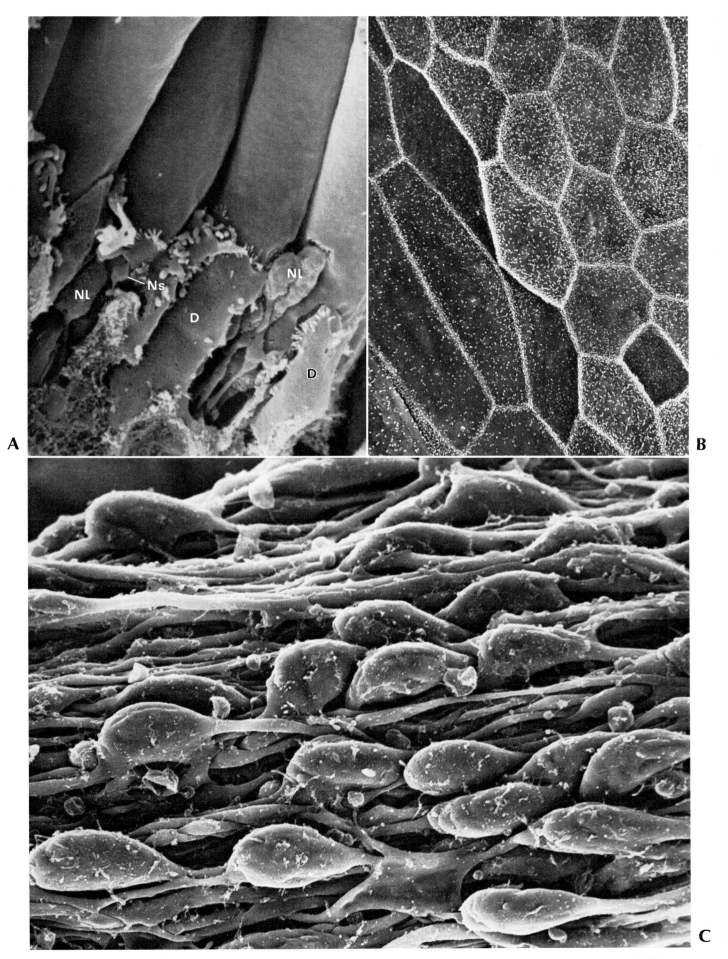

Eye

The eyeball is protected by a strong capsule of collagen fibers. Its posterior portion is the *sclera*, while its anterior, transparent portion is the *cornea*. The front of the cornea is covered by a stratified *epithelium* and its back by a thin *endothelium*. One of the earliest SEM applications to the biomedical field was the visualization of the rabbit corneal epithelium and endothelium by Blümcke and Morgenroth, Jr. (1967). The corneal epithelium later was described as composed of dark cells covered by numerous short microvilli and light cells possessing fewer and longer microvilli (Pfister, 1973). We conceive, however, that the bright appearance of the cells is largely due to the edge effect of the microvilli and becomes much less conspicuous by improved methods of specimen preparation.

The *iris* is composed, from the front to the back, of the stroma iridis, stratum pigmenti iridis and pars iridica retinae. The sphincter and dilater muscles are located in and behind the stroma and cause the subtle movement of the iris. The *stroma of the iris* is a collagen feltwork. SEM studies (Hansson, 1970 e; Dieterich *et al.*, 1971) confirmed the view reached by TEM investigators that flat fibroblasts incompletely cover the anterior surface of the stroma. Earlier histologists erroneously called this fibroblast layer the endothelium of the anterior chamber. The posterior surface of the iris, on the other hand, is continuously covered by the cells of the pars iridica retinae. They are polygonal cells with numerous microprojections (Hansson, 1970 e).

The study of the *lens* has been much benefited by the SEM since its early application by Tanaka (1969) and Hansson (1970 c) to the visualization of the interlocking patterns of lens fibers (Dickson and Crock, 1972; Harding *et al.*, 1976; Hollenberg *et al.*, 1976). The lens fibers are extended epithelial cells; many of them have lost their nuclei. They are flattened hexagons in cross section. The ridges of the fibers are provided with complicated microprojections and indentations with which they are interdigitated with adjacent fibers (Tanaka and Iino, 1967). The facets of the fibers may also possess parietal microprocesses and microplicae (Yamasaki, 1970). The morphology of lens fibers markedly differs among animals and its phylogenetical evaluation provides an interesting field of study.

The *retina* as the photoreceptive tissue seems the most interesting subject to be challenged by the SEM. The low magnification SEM images of adequately fractured retina (as shown in the atlas by Kessel and Kardon, 1979) are useful for the analysis of the neuronal chains and their supporting elements in the retina. With higher magnification, the fine surface structure of the *visual cells*, rods and cones, have been observed in the rat (Hansson, 1970), newt (Dickson and Hollenberg, 1971) and bullfrog (Steinberg, 1973). For other layers of the retina, including the pigment epithelium, SEM studies by Hansson (1970 a, b), Steinberg (1973) and Breipohl *et al.* (1973) are available.

The three-dimensional anatomy of the *iridocorneal angle* and the *Schlemm's canal* is important for the study of the mechanism and treatment of glaucoma. The SEM studies by Hansson and Jerndal (1971) and by Worthen (1972 a, b) are referred to concerning this subject. The delicate structures of the ciliary body (Krey, 1974; Davanger and Pedersen, 1978) and of the zonular fibers (Hansson, 1970 d) have been demonstrated by SEM.

Plate IX-5   Corneal Epithelium.   Bovine.

The substance of the cornea is a dense plate of strong collagen fibers. Its anterior surface is covered by a stratified epithelium composed of 5 or 6 cell layers. The micrograph shows the polygonal superficial cells of the *corneal epithelium*. Some cells are densely covered by a characteristic pattern of microplicae, whereas others are studded with granular microprojections.

X 6,900

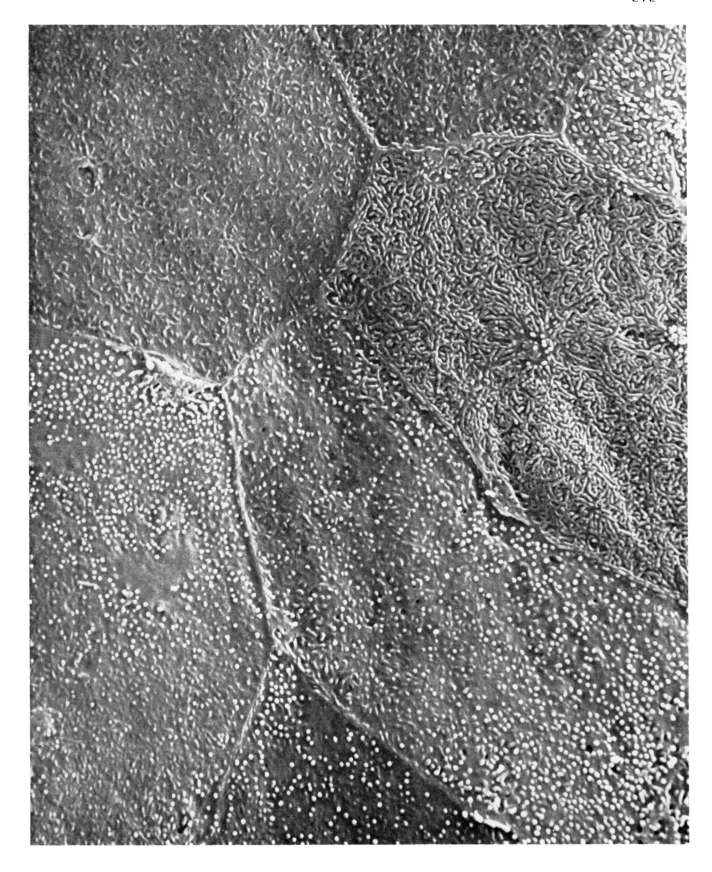

Plate IX-6   Corneal Endothelium.   Bovine.

The posterior surface of the cornea is covered by a thin endothelium. This endothelium faces the aqueous humor filling the anterior chamber of the eye.

A is an overview of the *endothelial cells.* They are flat in surface with very few microprojections. The cell margin only shows undulated elevations, thus the pentagonal or hexagonal cell boundaries are evident. Nuclear elevations are hardly visible as the nuclei of the cells are very flat.

B is a closer view of the cell boundaries of the corneal endothelium. Flat, tongue-like cell processes are interdigitated between adjacent cells and partly overlapped.

A: X 3,200,   B: X 23,000

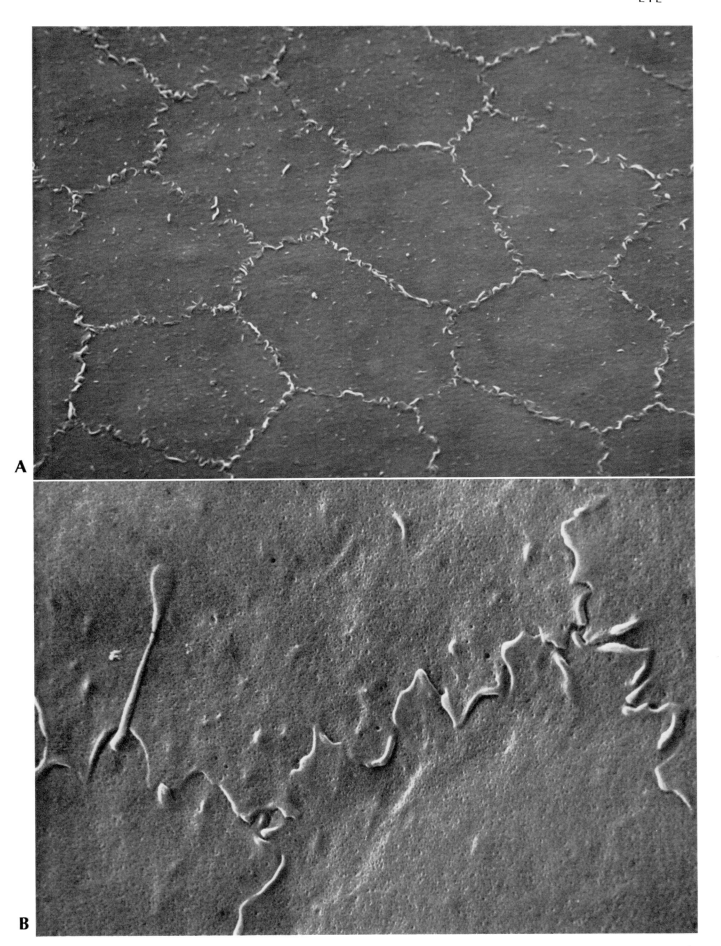

Plate IX-7 Iris. Rabbit.

The main, anterior portion of the iris is the *stroma iridis* (page 268) which is composed of loosely woven collagen fibers. The anterior surface of the stroma is incompletely covered by flattened fibroblasts of stellate shape. These cells irregularly overlap, leaving wide spaces where the collagen fibers are exposed to the aqueous liquor filling the anterior chamber.

X 6,000

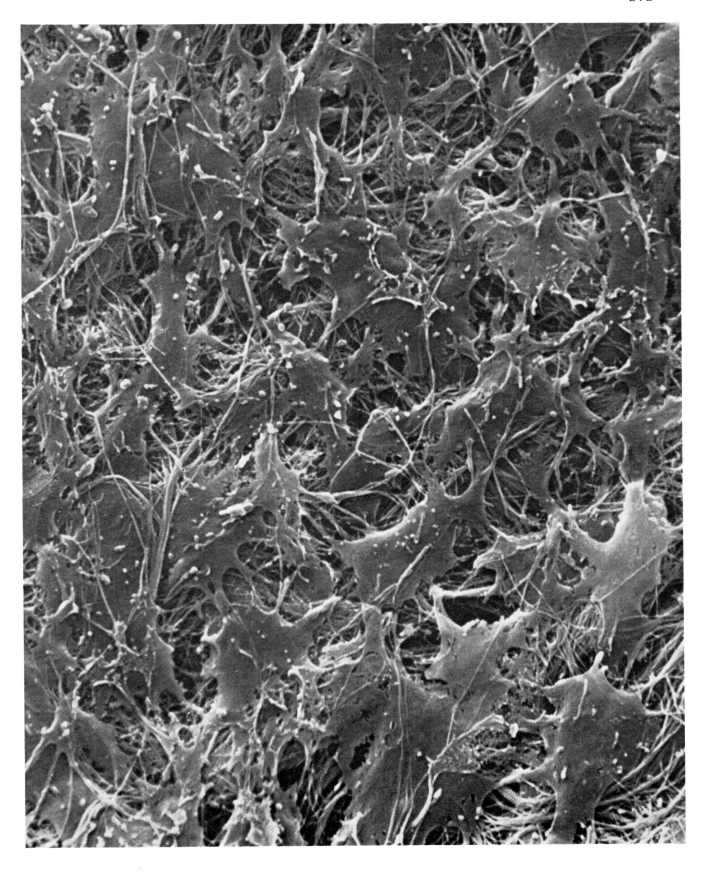

Plate IX-8
A. Fibroblast on the Stroma of the Iris. Rabbit.

A closer view of the *fibroblasts* covering the stroma iridis is demonstrated. The cells extend flat and tapering processes to connect with adjacent cells. At the bottom right the cell connection is clearly seen.

Larger and smaller bundles of collagen filaments run in different directions. Some of the filaments are seen on the surface of the cells.

X 22,000

B. Melanocytes in the Iris. Bovine.

This micrograph was obtained after the iris was broken under the dissection microscope. Deeper portions of the *stroma* were thus exposed and a few *melanocytes* of elongated shapes were demonstrated. The elypsoid granules filling up the cells are melanosomes.

The color of the iris which shows characteristic differences among races depends on the number, location and melanin contents of these cells.

X 9,200

Plate IX-9   Lens Fibers.   Bovine.

The lens was fractured and the lens fibers reveal their surface structures. Besides the large undulations on the lateral edge of the fibers, one may notice small knob-like projections (arrows). These structures represent the zippers and buttons with which the adjacent lens fibers are tightly connected.

   X 17,000

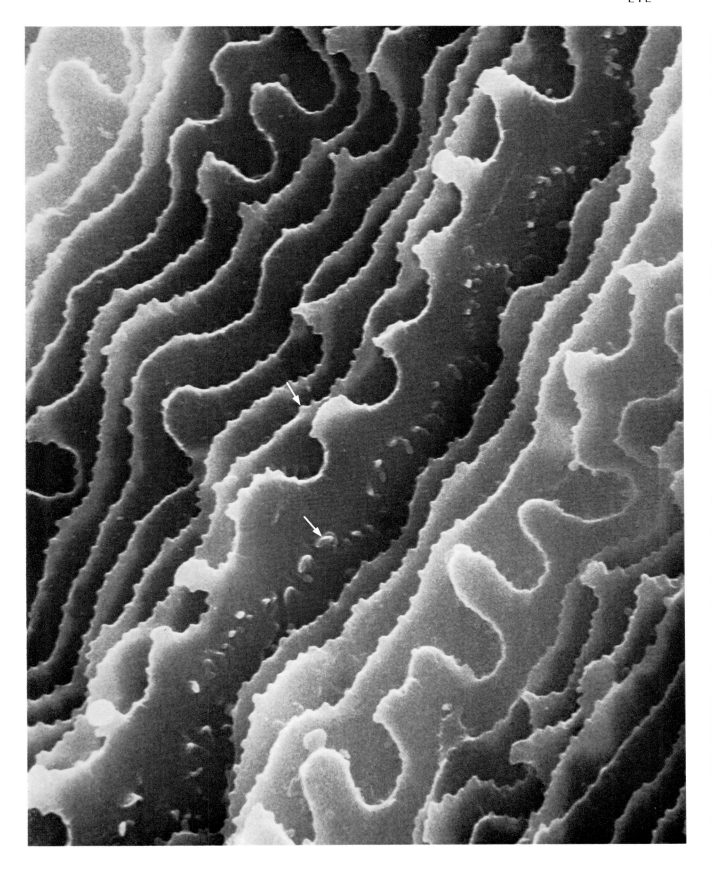

Plate IX-10   Retinal Rods and Cones, Overview.   Rhesus Monkey.

The pigment epithelium was peeled off and the *rods* and *cones* of retina are seen from the corneal side. Outer segments of the rods were partly removed together with the pigment epithelium and rods are more or less markedly shortened, so that the population of cones (arrows) is clearly seen, which otherwise are shorter than the rods and hardly visible.

   X 1,300

(Courtesy of Dr. Taeko Masutani, Department of Anatomy, Fukuoka University School of Medicine).

Plate IX-11   Retina.   Rhesus Monkey.

This micrograph shows an overview of fractured retina. In the *layer of rods and cones* (RC), slender rods occupy the great majority of visual segments and only a few cones (arrows) are seen.

The base of the rods and cones is marked by *outer limiting membrane* which is formed by attachment of processes of Müller's supporting cells.

Beneath the membrane is *outer granular layer* (OG) consisting of the nuclei-containing cell bodies of visual cells. This layer is followed by outer plexiform layer (OP), inner granular layer (IG), inner plexiform layer (IP) and ganglion cell layer (GC).

X 3,000

(Courtesy of Dr. Taeko Masutani, Department of Anatomy, Fukuoka University School of Medicine).

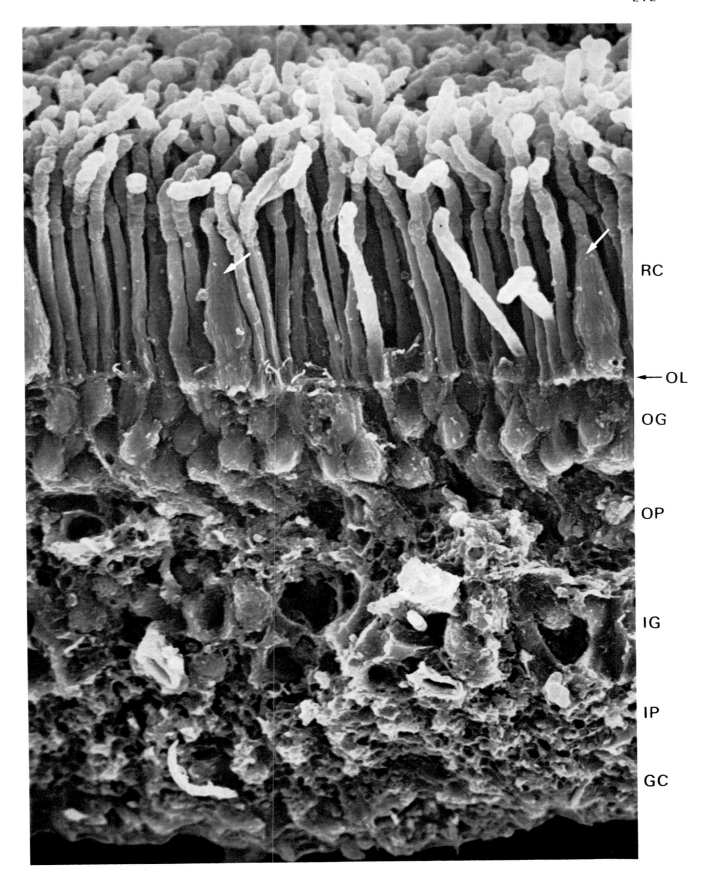

RC

OL

OG

OP

IG

IP

GC

Plate IX-12
A. Retinal Rods, Overview.   Bovine.

The pigment epithelium was peeled off and numerous *rods* were exposed. The bovine rods are columnar in shape, smooth in surface and possess a blunt end.

Some cells of the pigment epithelium are seen with their round granules. The pigment epithelial cells not only provide a dark background for the photoreceptor layer but also are involved in the synthesis of rhodopsin and in the turnover of the rods and cones by phagocytosing them bit by bit from their ends.

X 6,400

B. Layer of Visual Cells.   Frog.

The visual cells of the retina consist of *rod cells* and *cone cells.* Both cells possess their nucleated portion or the *inner segment* inside and their photoreceptor portion or the *outer segment* outside. The layer of the nucleated segment is called the *outer granular layer* (G), whereas the photoreceptor layer, the *layer of rods and cones* (RC). The boundary of both layers appears as the *external limiting membrane* (L) which is formed by attachment of the foot-like processes of Müller's supporting cells to the visual cells.

In the frog the cones (C) are much smaller than the rods (R) and their nuclei are located lower than those of the rods.

X 7,700

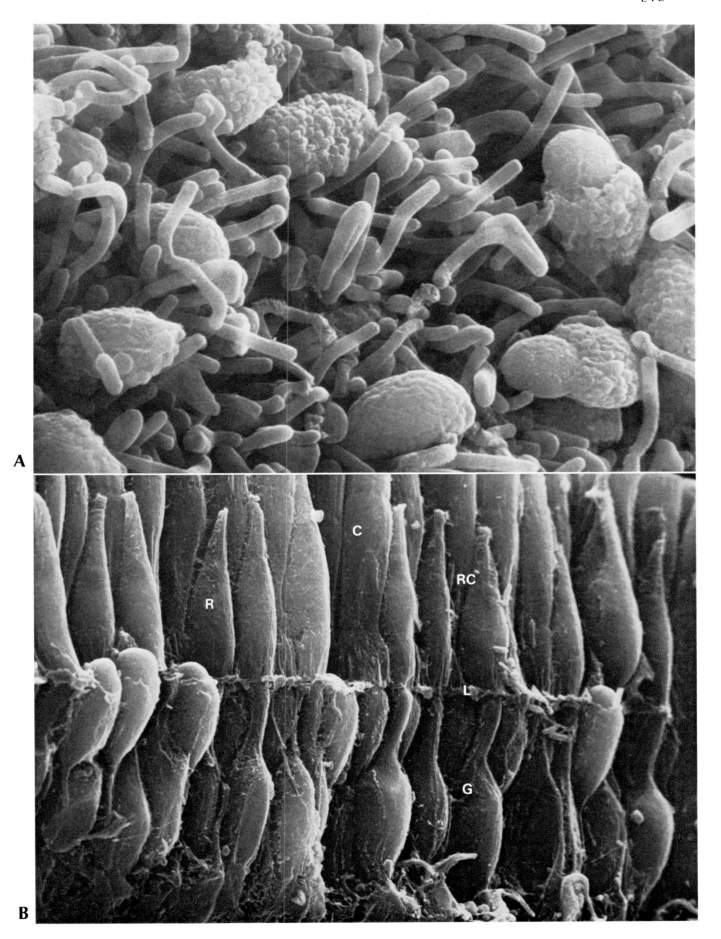

Plate IX-13   Cones and Rods.   Frog Retina.

Cones and rods are the outer segments of visual cells and essentially consist of piled up disks which represent the photoreceptor sites of the cells.

A shows two *cones* standing with the much taller rods. Deep invaginations occur crosswise which are involved in the formation of the disks (Dunn, 1973). A few slender processes longitudinally extend from the inner segment cytoplasm.

B demonstrates the surface structure of *rods*. Straight grooves run longitudinally and receive thin cytoplasmic processes ascending from the cell body or inner segment.

A: × 55,000,   B: × 74,000

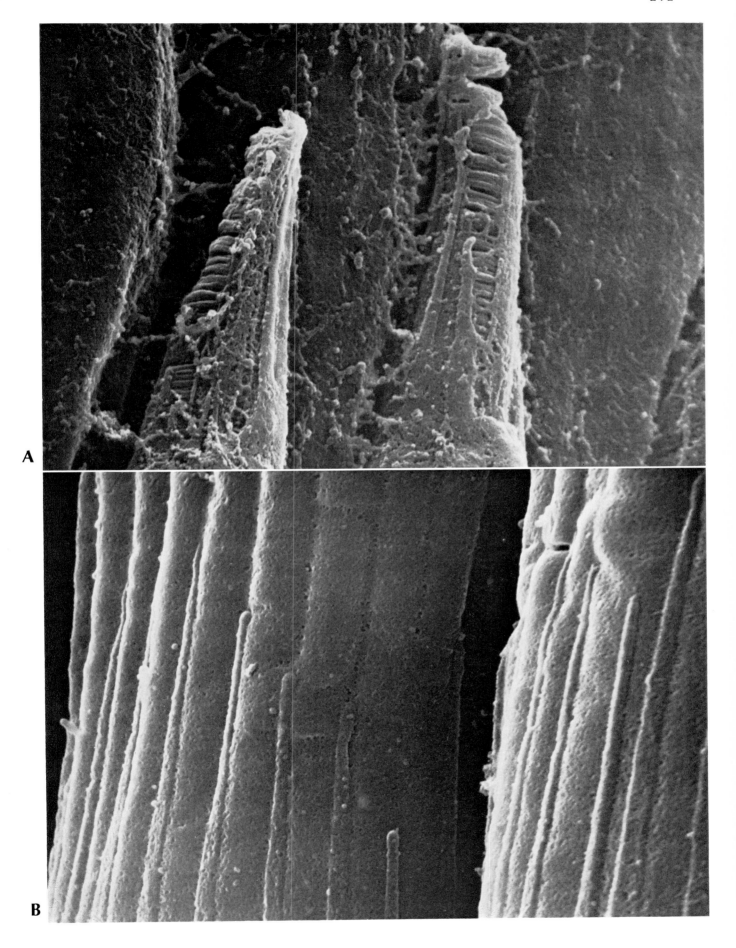

Olfactory Epithelium

The chemical stimuli in the inspired air are perceived by a specialized neuron-like cell (paraneuron) called the *olfactory cell* whose axonal process transmits the excitation to the brain to cause the smell sensation. The olfactory cells are distributed in the olfactory epithelium which occupies the upper and posterior corner of the nasal cavity.

The olfactory epithelium consists of three kinds of cells: *olfactory, sustentacular* and *basal cell.* The last is hidden from the surface and is believed to replace the other cells which undergo degeneration by damages and according to their life span.

Very little is known about the nature and function of the olfactory epithelium. For example: Are the olfactory cilia moving? Are they changing in number and length? What is the cell cycle of olfactory cell? What are the morphological changes in olfaction? Are the sustentacular cells only to sustain the olfactory cells? And if not, what are their activities? Studies in this field would seem to be accelerated by the application of SEM.

Andrews (1974) succeeded in showing the SEM view of olfactory mucosa in the rat. A recent TEM study by Saini and Breipohl (1976) in the monkey indicates conspicuous sex difference of mammalian olfactory cells, while another study by SEM from the same research group (Breipohl and Fernández, 1977) in the chicken suggests a promising field in the phylogenetic and ecological aspects of olfaction.

Plate IX-14   Olfactory Epithelium.   Human.

The *olfactory cells* project into the nasal cavity with a rounded head called the *olfactory vesicle,* which radiates columnar processes or *olfactory cilia.* This micrograph shows over a dozen of the olfactory vesicles, each projecting 15–18 cilia.

The *sustentacular cells* are covered by long and tortuous *microvilli* which, entangled by mucous substances, appear as a thick, reticular sheet embedding the olfactory cells.

X 12,000

(The specimen was kindly provided by Prof. M. Ohyama and Dr. I. Ohno, Department of Otorhinolaryngology at the Kagoshima University Medical School).

(The tissue was obtained from a cadaver (82-year-old male) fixed several hours after death by perfusion of 10% formalin through the femoral artery).

Plate IX-15
A. Olfactory Cell. Human.

An olfactory cell extends its *cilia* on the carpet of sustentacular cell microvilli. The proximal portion of the cilium tapers into a thinner distal portion which may be very long but tend to be broken during specimen preparation.
  X 31,000

B. Olfactory Cell. Guinea Pig.

The *olfactory bulb* in the dog is much larger in size than in the man.
  The *olfactory cilia* in this species are stouter and shorter in their proximal portion and much thinner in their distal portion (arrows) than in human. Small wart-like projections on the olfactory bulb are unknown in nature and in their possible relation to the cilia.
  X 23,000

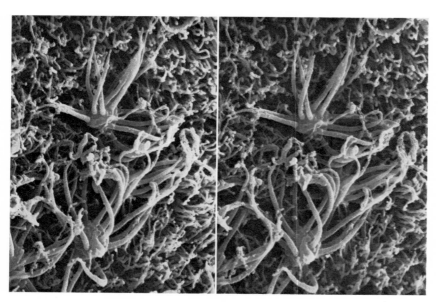

Fig. IX-3 Olfactory Epithelium in Stereo. Human.
Olfactory cells extending their cilia and long microvilli covering the sustentacular cells are demonstrated in stereo.
  X 8,000

(The specimens of plate IX-13A and Fig. IX-3 were kindly provided by Prof. M. Ohyama and Dr. I. Ohno, Department of Otorhinolaryngology, Kagoshima University Medical School. The tissue was obtained from 82-year-old male, see page 286).

Taste Bud

The taste bud has been studied under the SEM by only very few authors, and the observation has been mostly restricted to the taste pores (Arenberg *et al.*, 1969; Shimamura *et al.*, 1972; Arvidson, 1976). This atlas will demonstrate some new aspects of the cells forming the taste bud with special reference to the nerves entering the bud.

Plate IX-16   Taste Bud in Papilla foliata.   Rabbit.

A taste bud is fractured through the outer and inner pole. The taste cells forming the bud extend between both poles. A few *basal cells* (B) are located close to the inner pole and represent immature cells which can replace the long cells when they undergo degeneration.

The outer pole of the bud is characterized by the *microvilli* (M) of taste cells located in the *taste pore.* The inner pole is invaded with numerous *nerve fibers·* (N) which involve both afferent and efferent elements.

E indicates the ordinary layer of stratified epithelial cells surrounding the taste bud.   LP: lamina propria mucosae.

X 3,800

(Plate IX-16—18: Specimens of rabbit taste buds were prepared by Dr. K. Toyoshima, Department of Anatomy, Kyushu Dental College.)

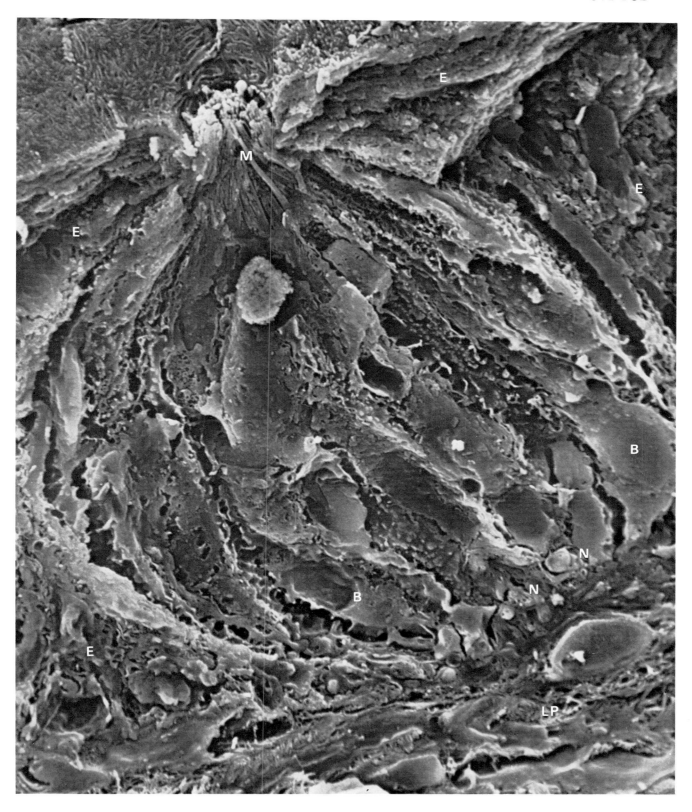

Plate IX-17   Taste Pore and Taste Hairs.   Rabbit, Papilla foliata.

The chemical stimuli of foodstuff are perceived by taste hairs which are the long microvilli of taste or gustatory cells.

A demonstrates a *taste pore* formed by the stratified epithelium (E) of the tongue whose characteristic surface relief may attract attention. The pore contains a dense tuft of *taste hairs*, which, by fracture, can be traced down to the surface of the taste bud (B). The conspicuous undulation of the bud cell surface is noteworthy.

B is a closer view of the taste hairs. Among the slender microvilli, a few stub-like cell processes are seen (arrows). The round bodies on the microvilli are unknown in nature. They may possibly be swollen portions of the cell processes.

A: X 10,000,   B: X 21,000

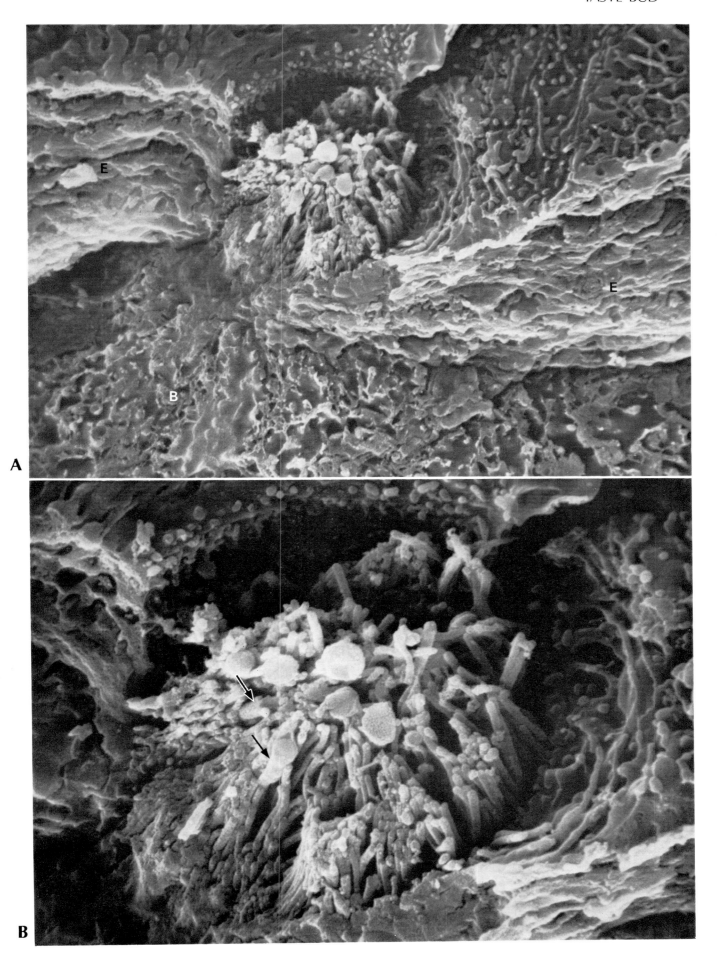

Plate IX-18   Nerves to Taste Bud. Rabbit, Papilla foliata.

Nerves enter the taste bud through its interior pole and, after winding round the taste cells, terminate on their surface forming synapses. At present the afferent and efferent nerves cannot be distinguished under the SEM.

A demonstrates the lower half of a fractured taste bud. The cells forming the taste bud partly reveal their undulated surfaces and partly their fractured cytoplasm.

Nerves (N) are identified here and there, partly accompanied by the thin sheath of Schwann cell cytoplasm (S). A groove (*) may be seen on the taste cell surface after a nerve has been removed.

B shows the nerves in a taste bud more closely. In the middle, three nerve fibers (N) embraced by attenuated Schwann cell (S) cytoplasm are fractured crosswise. At the bottom right, winding nerves (N) are exposed and their swellings (*) are tightly invaginated into the taste cell, apparently forming synapses.

A: X 8,800,   B: X 35,000

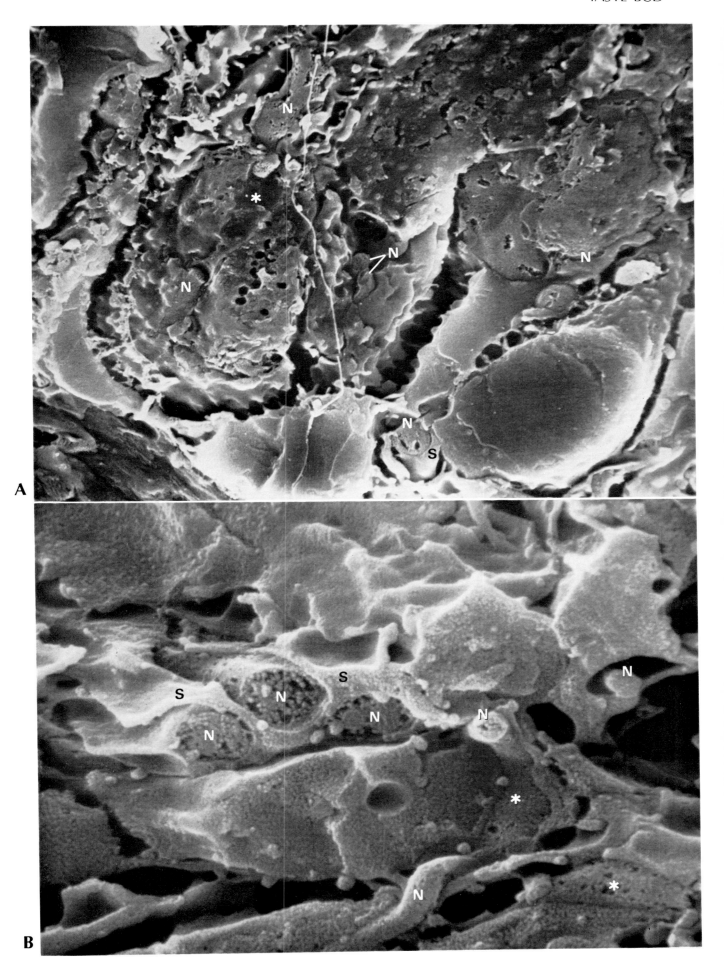

A

B

References

Andrews, P. M.: A scanning electron microscopic study of the extrapulmonary respiratory tract. *Amer. J. Anat.* **139**: 399—424 (1974).

Arenberg, I.K., W.F. Marovitz and A.P. Mackenzie: Scanning electron microscopy of the outer taste pore from a human tongue. *Proc. 27th Annu. Electron Microsc. Soc. Amer.* **3**: 36—37 (1969).

Arvidson, K.: Scanning electron microscopy of fungiform papillae on the tongue of man and monkey. *Acta Otolaryngol.* **81**: 496—502 (1976).

Blümcke, S. and K. Morgenroth, Jr.: The stereo ultrastructure of the external and internal surface of the cornea. *J. Ultrastr. Res.* **18**: 502—518 (1967).

Bredberg, G.: The innervation of the organ of Corti. A scanning electron microscopic study. *Acta oto-laryngol. (Stockh.)* **83**: 71—78 (1977).

Breipohl, W., N. Bornfeld, G.J. Gijvank, H. Laugwitz and M. Ptautsch: Scanning electron microscopy of the retinal pigment epithelium in chick embryos and chicks. *Z. Zellforsch.* **146**: 543—552 (1973).

Breipohl, W. and M. Fernández: Scanning electron microscopic investigations of olfactory epithelium in the chick embryo. *Cell Tiss. Res.* **183**: 105—114 (1977).

Davanger, M. and O.Ö. Pedersen: The ciriary body and the suspension of the lens in rabbits. A scanning electron microscopy study. *Acta ophthalmol.* **56**: 127—138 (1978).

Dickson, D.H. and G.W. Crock: Interlocking patterns of primate lens fibers. *Invest. Ophthalmol.* **11**: 809—815 (1972).

Dickson, D.H. and M.J. Hollenberg: The fine structure of the pigment epithelium and photorecepter cells of the newt, *Triturus viridescens dorsalis* (Refinesque). *J. Morphol.* **135**: 389—432 (1971).

Dieterich, C.E., R. Witmer and H.E. Franz: Iris und Kammer-Zirkulation. Morphologische Analysen der Oberflächenstrukturen der menschlichen Iris. *Albrecht v. Graefes Arch. klin. exp. Ophthalmol.* **182**: 321—340 (1971).

Dunn, R. F.: The ultrastructure of the vertebrate retina. In: (ed. by) I. Friedman: The ultrastructure of sensory organs. North-Holland Pub. Co., Amsterdam-London, 1973 (p. 153—265).

Hansson, H.-A.: Scanning electron microscopy of the rat retina. *Z. Zellforsch.*, **107**: 23—44 (1970a).

Hansson, H.-A.: Scanning electron microscopic studies on the synaptic bodies in the rat retina. *Z. Zellforsch.* **107**: 45—53 (1970b).

Hansson, H.-A.: Scanning electron microscopy of the lens of the adult rat. *Z. Zellforsch.* **107**: 187—198 (1970c).

Hansson, H.-A.: Scanning electron microscopy of the zonular fibers in the rat eye. *Z. Zellforsch.* **107**: 199—209 (1970d).

Hansson, H.-A.: Ultrastructure of the surface of the iris in the rat eye. *Z. Zellforsch.* **110**: 192—203 (1970e).

Hansson, H.-A and T. Jerndal: Scanning electron microscopic studies on the development of the iridocorneal angle in human eyes. *Invest. Ophthalmol.* **10**: 252—265 (1971).

Harding, C.V., S. Susan and H. Murphy: Scanning electron microscopy of the adult rabbit lens. *Ophthalmic. Res.* **8**: 443—455 (1976).

Held, H.: Die Cochlea der Sauger und der Vogel. Ihre Entwicklung und ihr Bau. In: Handbuch der normalen und pathologischen Physiologie. 12, Springer, Berlin, 1926.

Hollenberg, M.J., J.P.H. Wyse and B.J. Lewis: Surface morphology of lens fibers from eyes of normal and microphthalmic (Browman) rats. *Cell Tiss. Res.* **167**: 425—438 (1976).

Hoshino, T.: Relationship of the tectorial membrane to the organ of Corti. A scanning electron microscope study of cats and guinea pigs. *Arch. histol. jap.* **37**: 25—39 (1974).

Hoshino, T.: The structure of the cochlea; especially techtorial membrane-inner hair cell relation, Deiter's cells and basilar membrane cells. (In Japanese) *Jibi-Inkoka* **47**: 951—958 (1975).

Hoshino, T.: Contact between the tectorial membrane and the cochlear sensory hairs in the human and the monkey. *Arch. Oto-Rhino-Laryng.* **217**: 53—60 (1977).

Hunter-Duvar, I. M.: Morphology of the normal and the acoustically damaged cochlea. In: (ed. by) O. Johari and R. P. Becker: Scanning Electron Microscopy/1977 II. IIT Research Institute, Chicago, 1977 (p. 421—428).

Hunter-Duvar, I. M. and R. J. Mount: The organ of Corti following ototoxic antibiotic treatment. *Ibid./1978. II.* Scanning Electron Microscopy, Inc., AMF O'Hare, 1978 (p. 423—430).

Kessel, R. G. and R. H. Kardon: Tissues and Organs; A Textatlas of Scanning Electron Microscopy. W. H. Freeman and Company, San Francisco, 1979.

Krey, H.: Zur Raster-Elektronenmikroskopie der Pars plicata des menschlichen Ciliarkörpers. *Albrecht v. Graefes Arch. klin. exp. Ophthalmol.* **191**: 127—137 (1974).

Leake-Jones, P. A. and M. C. Vivion: Cochlear pathology in cats following administration of neomycin sulfate. In: (ed. by) R. P. Becker and O. Johari: Scanning Electron Microscopy/1979 III. Scanning Electron Microscopy, Inc., AMF O'Hare, 1979 (p. 983—991).

Lim, D. J.: Three dimensional observation of the inner ear with the scanning electrone microscope. *Acta otolaryngol., Suppl.* **255**: 1—38 (1970).

Lim, D. J.: Ultrastructural cochlear changes following acoustic hyperstimulation and ototoxicity. *Ann. Otol. Rhinol. Laryngol.* **85**: 740—751 (1976).

Lim, D. J. and W. C. Lane: Cochlear sensory epithelium: A scanning electron microscopic observation. *Ann. Otol. Rhinol. Laryngol.* **78**: 827—841 (1969).

Lindeman, H. H. and G. Bredberg: Scanning electron microscopy of the organ of Corti after intense auditory stimulation: Effects on stereocilia and cuticular surface of hair cells. *Arch. klin. exp. Ohr.- Nas.- Kehlk.-Heilk.* **203**: 1—15 (1972).

Pfister, R. R.: The normal surface of corneal epithelium; a scanning electron microscopic study. *Invest. Ophthalmol.* **12**: 654—668 (1973).

Ross, M. D.: The tectorial membrane of the rat. *Amer. J. Anat.* **139**: 449—482 (1974).

Saini, K. D. and W. Breipohl: Surface morphology in the olfactory epithelium of normal male and female rhesus monkeys. *Amer. J. Anat.* **147**: 433—446 (1976).

Shimamura, A., J. Tokunaga and H. Toh: Scanning electron microscopic observations on the taste pores and taste hairs in rabbit gustatory papillae. *Arch. histol. jap.* **34**: 51—60 (1972).

Shinozaki, N. and M. Miyoshi: Scanning electron microscopy on the innervation of Corti's organ. *Arch. histol. jap.* (in press).

Steinberg, R. H.: Scanning electron microscopy of the

Bullfrog's retina and pigment epithelium. *Z. Zellforsch.* **143**: 451–463 (1973).

Tanaka, K.: Darstellung der Linsenfasern von Fischen anhand von Abdrücken und mittels des Raster-Elektronenmikroskops. *Arch. histol. jap.* **30**: 233–246 (1969).

Tanaka, K. and A. Iino: Zur Frage der Verbindung der Linsenfasern im Rinderauge. *Z. Zellforsch.* **82**: 604–612 (1967).

Tanaka, T., N. Kosaka, T. Takiguchi, T. Aoki and S. Takahara: Observation on the cochlea with SEM. In: (ed. by) O. Johari and I. Corvin: Scanning Electron Microscopy/1973. IIT Research Institute, Chicago, 1973 (Part III, p. 427–434).

Theopold, H.-M.: Comparative surface studies of ototoxic effects of various aminoglycoside antibiotics on the organ of Corti in the guinea pig. A scanning electron microscopic study. *Acta otolaryngol.* **84**: 57–64 (1977).

Worthen, D. M.: Endothelial projection in Schlemm's canal. *Science* **175**: 561–562 (1972a).

Worthen, D. M.: Scanning electron microscopic study of the interior of Schlemm,'s canal in the human eye. *Amer. J. Ophthalmol.* **74**: 35–40 (1972b).

Yamasaki, H.: Lens fiber junction systems of reptiles and birds. *Arch. histol. jap.* **32**: 41–50 (1970).

CHAPTER X

ENDOCRINE GLANDS

Endocrine glands, except for the thyroid (*vide infra*), have been observed under the SEM by only a few authors. The cellular architectonics (Motta *et al.*, 1979) and the morphology of secretory granules have been the main subjects of study while observation of blood vascular casts taken from endocrine tissues have provided important information concerning the hormone transportation routes (Fujita and Murakami, 1973; Fujita and Murakami, 1974; Fujita *et al.*, 1976; Murakami, 1975).

Among divergent kinds of endocrine glands this atlas will concentrate on the thyroid, pituitary and pancreatic islet.

Thyroid

The thyroid follicles, when fractured, tend to expose their inner surface, the colloid which filled the follicles being removed. The apical aspect of follicle epithelial cells is covered by microvilli (Hansen and Skaaring, 1973; Bucher and Krstić, 1978) and frequent single cilium (Kobayashi, 1973) has been demonstrated. The observation by Ketelbant-Balasse *et al.* (1973) demonstrated spoon-shaped pseudopods on the follicle cell surface appearing after TSH stimulation. This structure apparently represents the tongue formation of cytoplasm for embracing and incorporating the colloid, i.e., thyroglobulin reabsorption by the follicle epithelial cells.

Plate X-1   Thyroid Follicles.   Dog.

One large and some smaller follicles are opened, while the colloid has been removed. The *follicle epithelial cells*, are domed and covered with *microvilli*. The occurrence of *single cilium* is clear at this magnification.

Parafollicular cells (P) which produce the peptide hormone, calcitonin, are clearly shown.
X 7,400

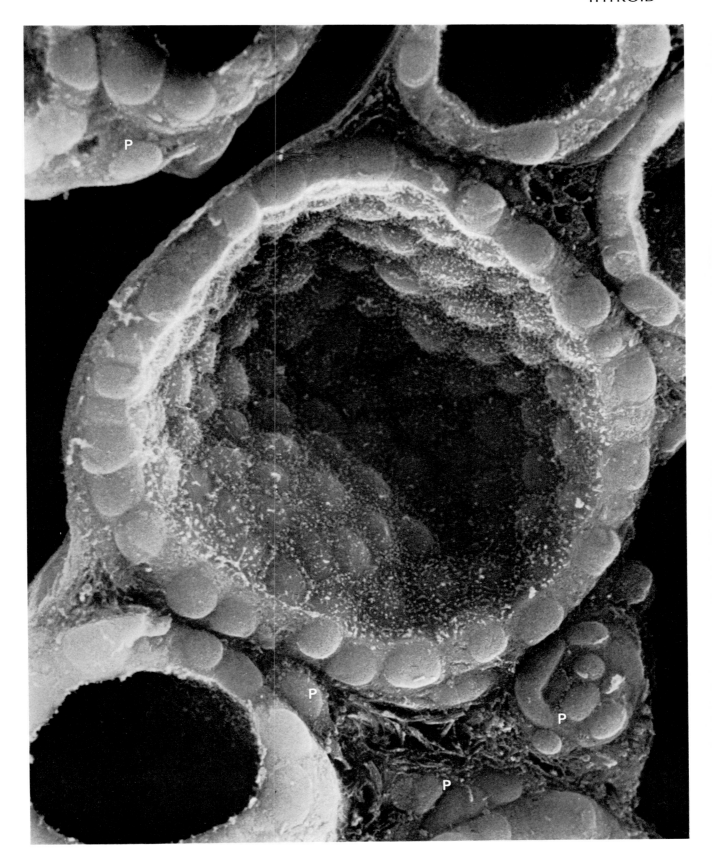

Plate X-2
A. Thyroid Follicle Epithelium. Dog.

Follicle epithelial cells are fractured showing three nuclei (N). The cytoplasmic matrix has been macerated in this specimen, and mitochondria (M) and endoplasmic reticula stand out.
   The luminal or apical cell surface (L) of the follicle epithelial cells is partly seen.
   X 8,300

B. Follicle Luminal Surface. Dog.

The luminal surface of the follicle epithelial cells is covered by microvilli projecting in irregular directions.
   Apocrine-like globules (arrows) may be seen on the surface of the cell (Kobayashi, 1973).
   X 16,500

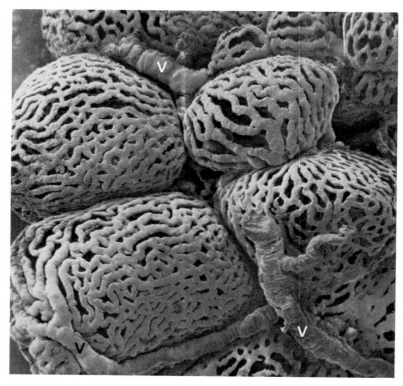

Fig. X-1 Vascular Cast of Thyroid.   Rhesus Monkey.
Each follicle is covered by a distinct layer of capillary netting which is drained by interfollicular veins (V).
   X 125

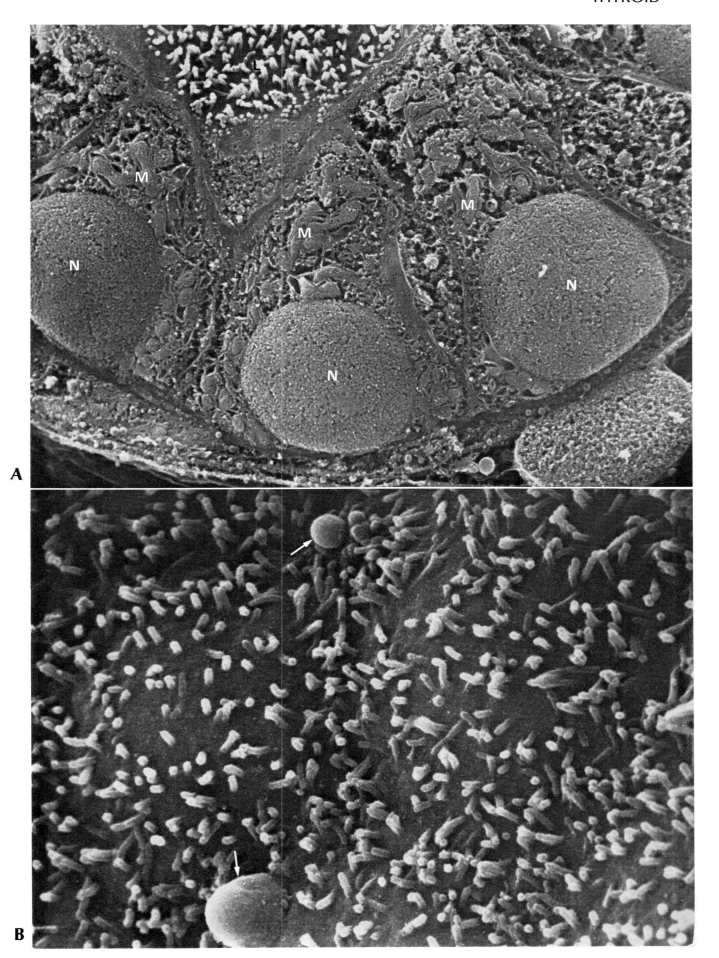

A

B

Pituitary Gland

Plate X-3   Anterior Pituitary or Adenohypophysis.   Rabbit.

Endocrine cells forming the pituitary anterior lobe have been fractured and secretory granules measuring 100—450 nm are exposed. As labels indicate, it is possible to identify the cell types (accordingly the cell products) under the SEM, on the basis of their cell and granule morphology established by the use of the TEM (Salazar, 1963; Young *et al.*, 1965; Kurosumi, 1968).

A demonstrates five kinds of granulated cells (STH, FSH, LTH, LH, TSH) and a follicular or agranular cell (FC). Special attention should be paid to the fact that the granules are conspicuously different in size within one and the same cell, which is difficult to evidence in thin section studies by TEM.

B   highlights STH and ACTH cells and includes some agranular or follicular cells (FC).

A: X 7,500,   B: X 6,300

(Plate X-3: Courtesy of Dr. T. Nakano, Department of Anatomy, Tottori University Medical School).

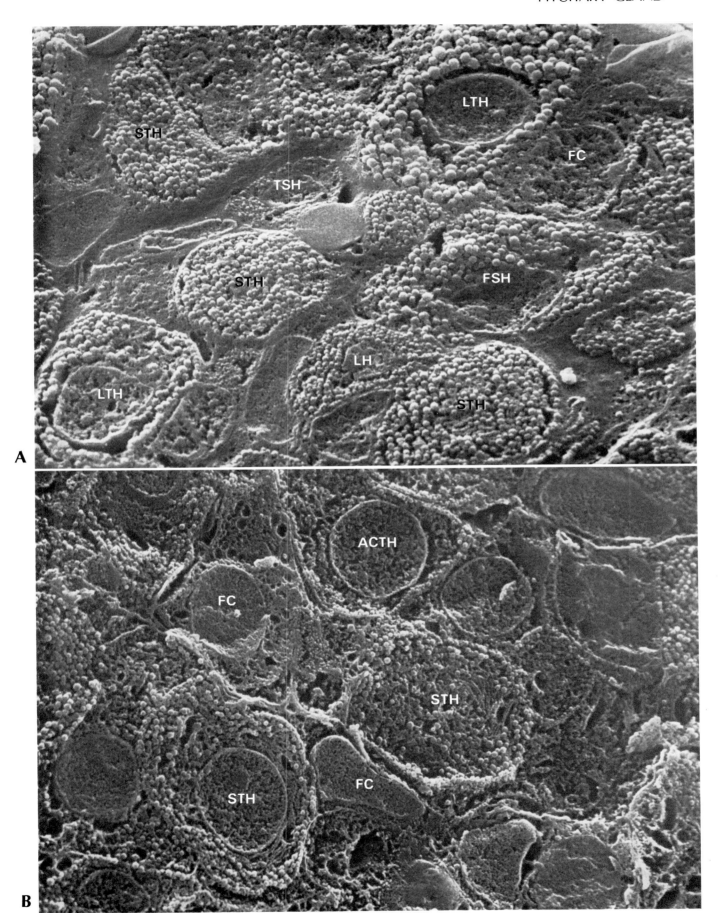

Plate X-4   Posterior Pituitary or Neurohypophysis.   Rabbit.

The pituitary posterior lobe comprises nerve endings containing secretory materials (oxytocin and vasopressin) derived from the hypothalamic nuclei.

A shows swollen portions of nerve fibers which contain a few secretory granules (arrows). Their diameter measures about 100 nm.

B demonstrates a few Herring's bodies, i.e., swollen parts of nerves filled with granules, which measure about 150 nm but again show a marked size variability within the same nerve. TEM images of rabbit neurohypophysis are available in Barer and Lederis (1966).

A: X 13,500,   B: X 18,000

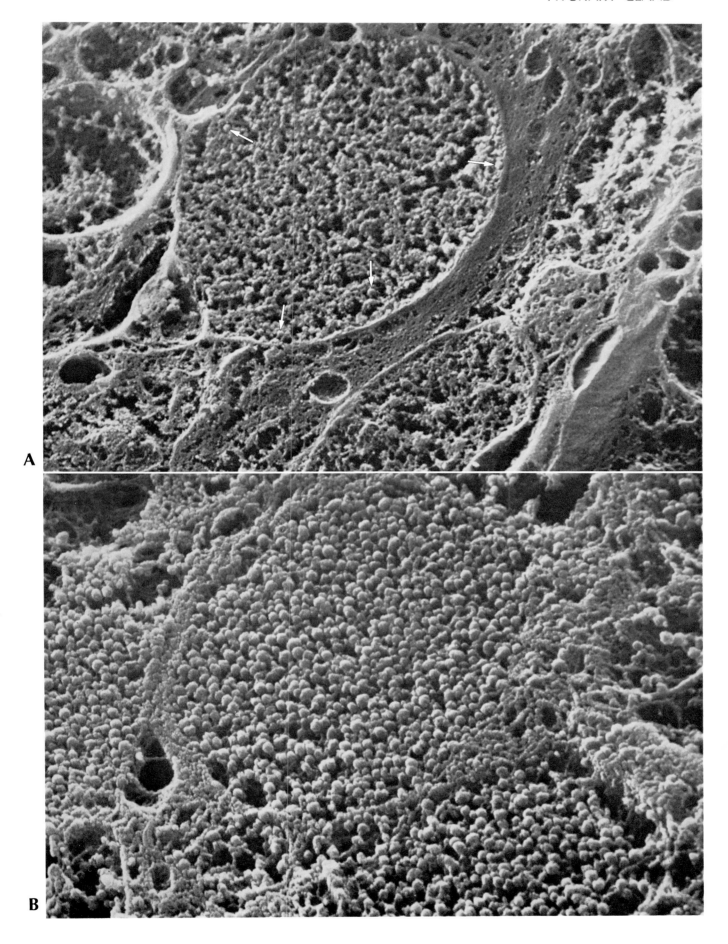

Pancreatic Islet

Plate X-5 Islet of Langerhans. Dog Pancreas.

The islet of Langerhans consists of polyhedral endocrine cells. A, B and D cells correspond to the secretion of glucagon, insulin and somatostatin, the B cells comprising the major cell type. The islet receives ample blood vessels which are sinusoidal capillaries.

A shows, on the right bottom, capillary endothelium (E) of pored structure. Note the small round indentations which correspond to the pores, as well as the microvilli on the luminal surface of the endothelium.

The *pericapillary space* (P) appears either empty as in **A** or partly occupied by collagen fibrils (Pc) as in **B**. Islet cells facing the pericapillary space may be provided with microvilli of different length (arrows).

The B cells are fractured and the core disks of their secretory granules are evident. The arrowheads indicate exocytotic release of the granules.

B further shows a *nerve bundle* (N) in the pericapillary space. The occurrence of nerve fibers and their terminals at this position is common in the dog and it has been proposed that they might release their secretions into the blood, i.e., they might be involved in neurosecretion in the pancreatic islet (Fujita and Kobayashi, 1979).

A: X 12,500,   B: X 16,000

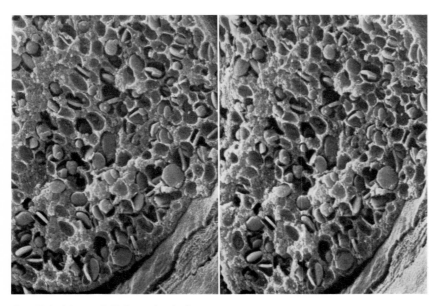

Fig. X-2 Islet B Cell Granules in Stereo.   Dog.
The granules consist of a vacuole and a disk-shaped core; insulin is known to be contained in the latter. Note the conspicuous size difference in both the vacuoles and cores and the occasional occurrence of double cores.
X 7,800

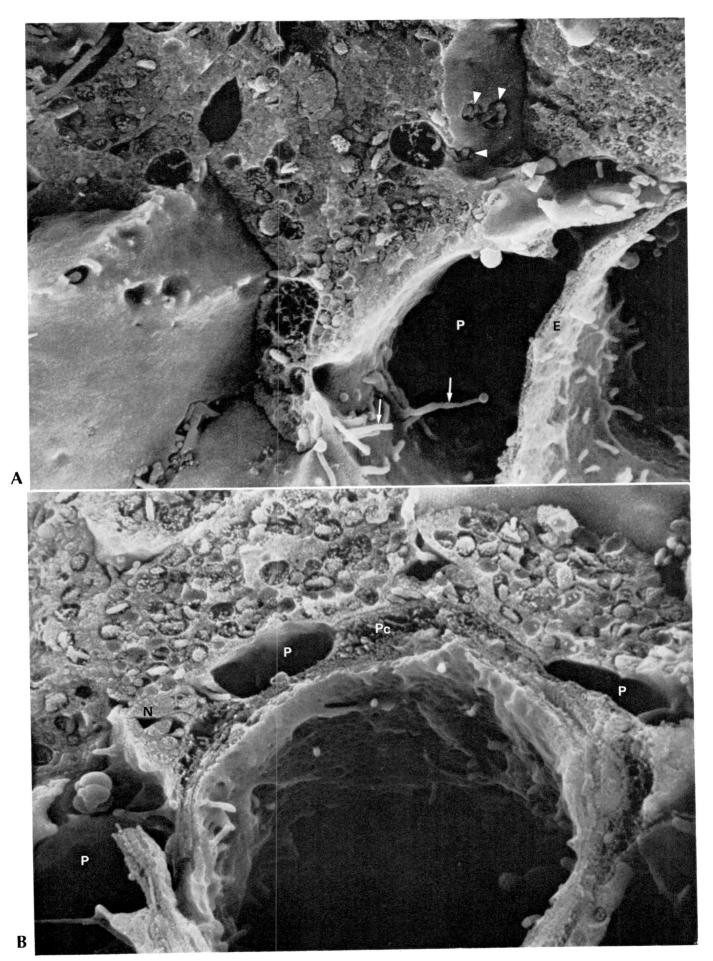

Plate X-6 Architecture of Islet Cells. Dog Pancreas.

A shows a sinusoidal capillary (C) and pericapillary spaces (P). B cells (B) are fractured showing their discoid granule cores.

In the upper left fracture occurred between the cell boundaries, and the flat facets of polyhedral cells are exposed. Along the edges of the cells, where usually three cells meet together as is indicated by TEM images of sections, there extends a *canalicule* (arrows) which contains numerous microvilli issuing from the cells concerned. This canalicule is connected to the pericapillary space and hypothetically represents an extension of the pericapillary face of the cell sensitive to chemical information from the blood, especially to glucose levels (Fujita *et al.*, 1981).

The exocytotic granule release (arrowheads) occurs on the flat part of the B cell surface.

B is a closer view of the *intercellular canalicule* (arrows). The cell edges of two adjacent B cells extend irregular microvilli which may possibly be the sites of glucoreceptors and other receptors. The flat cell surface shows holes connected to a granular substance which are believed to represent the processes of exocytotic granule release (arrowheads).

A: × 8,000, B: × 16,000

(Plate X-6: From T. Fujita, S. Kobayashi and Y. Serizawa *Biomed. Res.* 2, Suppl. : 115—118, 1981)

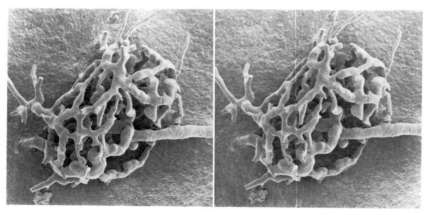

Fig. X-3 Resin Cast of Islet Blood Vessels. Dog.
An islet was dissected out from a vascular cast of dog pancreas. It is clearly seen that an arteriole (*vas afferens*) enters the capillary glomus of the islet, which is drained to the exocrine pancreas via radiating capillaries (*vasa efferentia*). The exocrine pancreas thus receives islet hormones in high concentrations (Fujita and Murakami, 1973; Fujita *et al.*, 1976; Ohtani and Fujita, 1980).
× 275

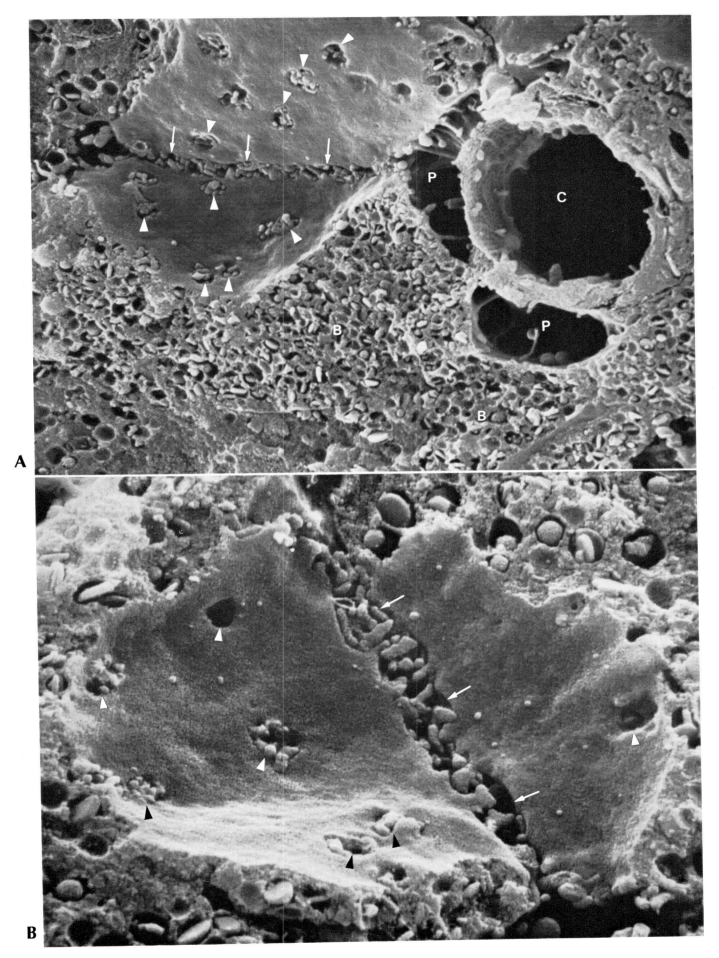

References

Bucher, O. and R. Krstić: Rasterelektronenmikroskopische Untersuchung der Oberflächenveränderungen von Schilddrüsenepithelzellen der Ratte nach 6- bis 48-stündiger Kälteeinwirkung. *Anat. Embryol.* **153**: 85—93 (1978).

Fujita, H. and T. Murakami: Scanning electron microscopy on the distribution of the minute blood vessels in the thyroid gland of the dog, rat and rhesus monkey. *Arch. histol. jap.* **36**: 181—188 (1974).

Fujita, T. and S. Kobayashi: Proposal of a neurosecretory system in the pancreas. An electron microscope study in the dog. *Arch. histol. jap.* **42**: 277—295 (1979).

Fujita, T., S. Kobayashi and Y. Serizawa: Canalicule system in the pancreatic islet of dog and mouse. A combined SEM and TEM study. *Biomed. Res.* **2**, Suppl. : 115—118 (1981).

Fujita, T. and T. Murakami: Microcirculation of monkey pancreas with special reference to the insulo-acinar portal system. A scanning electron microscope study of vascular casts. *Arch. histol. jap.* **35**: 255—263 (1973).

Fujita, T., Y. Yanatori and T. Murakami: Insulo-acinar axis, its vascular basis and its functional and morphological changes caused by CCK-PZ and caerulein. In: (ed. by) T. Fujita: Endocrine Gut and Pancreas. Elsevier, Amsterdam, 1976 (p. 347—357).

Hansen, J. and P. Skaaring: Scanning electron microscopy of normal rat thyroid. *Anat. Anz.* **134**: 177—185 (1973).

Ketelbant-Balasse, P., F. Rodesch, P. Neve and J. M. Pasteels: Scanning electron microscope observations of apical surfaces of dog thyroid cells. *Exp. Cell Res.* **79**: 111—119 (1973).

Kobayashi, S.: Rasterelektronenmikroskopische Untersuchungen der Shilddrüse. *Arch. histol. jap.* **36**: 107—117 (1973).

Kurosumi, K.: Functional classification of cell types of the anterior pituitary gland accomplished by electron microscopy. *Arch. histol. jap.* **29**: 329—362 (1968).

Motta, P., M. Muto and T. Fujita: Three dimensional organization of mammalian adrenal cortex. A scanning electron microscopic study. *Cell Tissue Res.* **196**: 23—28 (1979).

Murakami, T.: Pliable methacrylate casts of blood vessels: Use in a scanning electron microscope study of the microcirculation in rat hypophysis. *Arch. histol. jap.* **38**: 151—168 (1975).

Ohtani, O. and T. Fujita: Microcirculation of the pancreas with special reference to periductular circulation. A scanning electron microscope study of vascular casts. *Biomed. Res.* **1**: 130—140 (1980).

Salazar, H.: The pars distalis of the female rabbit hypophysis: An electron microscopic study. *Anat. Rec.* **147**: 469—497 (1963).

Young, B. A., C. L. Foster and E. Cameron: Some observations on the ultrastructure of the adenohypophysis of the rabbit. *J. Endocrinol.* **31**: 279—287 (1965).

CHAPTER **XI**

MUSCLES, NERVES AND BRAIN

This chapter consists of a rather miscellaneous collection of SEM images including skeletal muscle fibers, neuromuscular junctions, myelinated nerve fibers and some portions of the brain.

Muscles and Nerves

Plate XI-1
A. Skeletal Muscle Fibers. Rabbit Tongue.

The intrinsic muscles of the tongue are typical skeletal muscles. The micrograph shows obliquely fractured fibers from one of those muscles. Each muscle fiber is ensheathed by a *sarcolemma* (L), plasma membrane supported by a thin cytoplasmic layer. The major portion of the muscle fiber is filled with *muscle fibrils* which appear as parallel rods.
   X 1,750

B. Closer View of Muscle Fiber. Rabbit Tongue.

A portion of longitudinally fractured muscle fiber is shown. The *muscle fibrils* (F) of various thickness are now clear and rows of *mitochondria* (M) accompany them. Some mitochondria are fractured and reveal their cristae. The transverse striation on the myofibrils is formed by *T* or *transverse tubules* (T) strengthened by a membrane system of sarcoplasmic reticulum attached to them. The *sarcoplasmic reticulum* (S) is shrunken in this preparation and its extended saccules are seen only incompletely.
   X 12,500

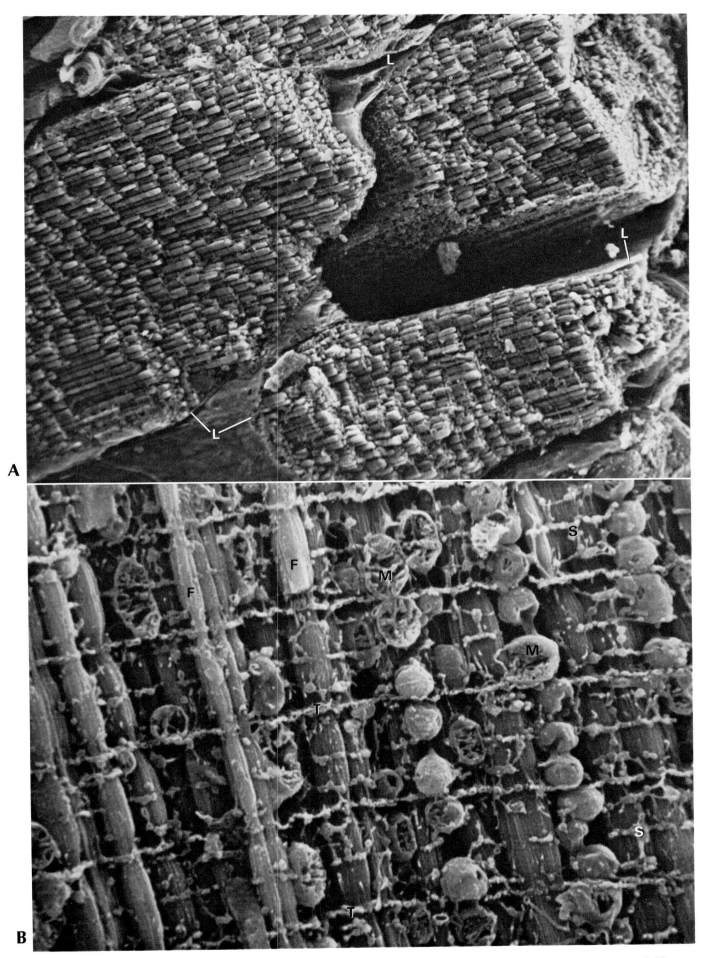

Plate XI-2   Sarcoplasmic Reticulum and T Tubule.   Frog Sartorius Muscle.

The excitation on the sarcolemma of the muscle fiber is immediately conducted deep into the muscle fiber via the system of *transverse* or *T tubules*, which, in their turn, transmit the excitation to another membrane system, the *sarcoplasmic reticulum* which invests the myofibrils.

   The micrograph shows a surface view of a *myofibril* after a fracture which just grazes the membrane systems. Thus, a complex called the *triad*, formed by a T tubule and two sarcoplasmic reticula, is seen in two places in this micrograph. In the triad a T tubule (T) is sandwiched by the *terminal cisternae* (TC) of the sarcoplasmic reticula. The central portions of the sarcoplasmic reticula (S) have been removed leaving only a small part of them.

   There are seen three myofibrils in this micrograph. Two kinds of myofilaments, actin and myosin filaments, cannot clearly be identified.

   X 65,000

(Courtesy of Dr. H. Sawada, Department of Anatomy, University of Tokyo Faculty of Medicine. Reproduced from Sawada *et al.*: *Tissue and Cell* **10**: 179–190, 1978).

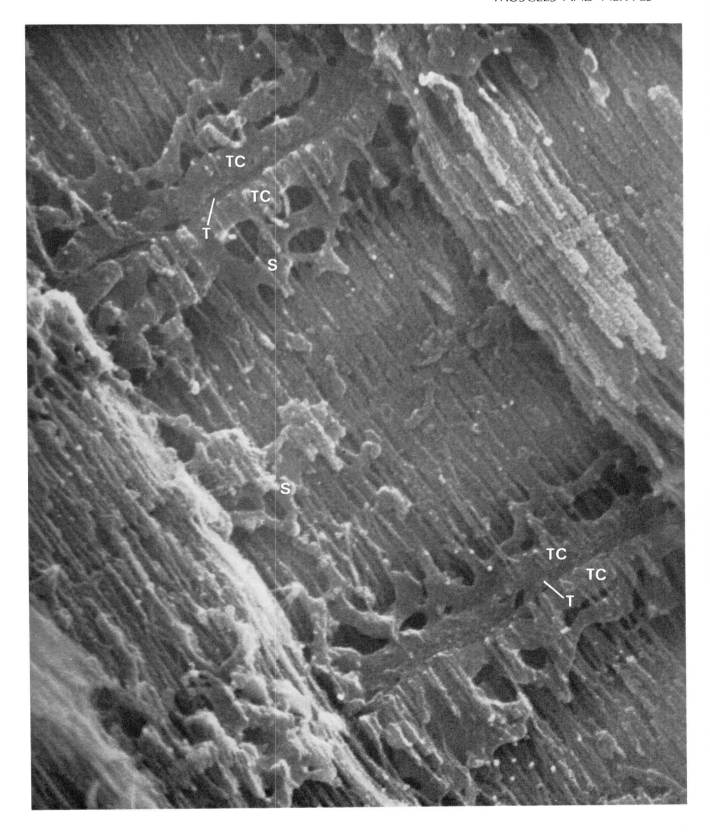

Plate XI-3
A. Neuromuscular Junction. Chinese Hamster.

Muscular tissue was treated for 30 min with 8 N HCl at 60°C. This procedure removed collagen fibers and basement membranes from the tissue and the surfaces of cellular structures were denuded.

The micrograph shows a bundle of motor nerves (N) which end in two *motor endplates* (E). The terminal portion of the nerve fibers are arborized into anastomosing and twisting endfeet which are accepted by corresponding folds on the side of the muscle. Synaptic vesicles containing acetylcholine are gathered in these endfeet. A *Schwann cell* (S) may be seen associated with the neuromuscular junction and may be called a *teloglial cell.*

A blood capillary (C) is seen with one of its pericytes (P).

X 1,650

B. Synaptic Folds of Neuromuscular Junction. Chinese Hamster.

Occasionally the endplate of the neuromuscular junction may be removed during the HCl treatment above mentioned. Then, twisted grooves are exposed which are the *synaptic folds* gathered in an elongate area. In the bottom of the folds one can see lamellar slits deeply extending into the muscle fiber. These are called *subsynaptic* or *secondary folds.* These folds apparently increase the cell surface area receiving the synaptic control of the nerve.

Arrows indicate the *sarcoplasmic eminence* which is caused by an accumulation of nuclei and mitochondria.

For further details of these findings of neuromuscular junction, see Uehara *et al.* (1981).

X 3,000

(Plate XI-3: Courtesy of Prof. Y. Uehara, Deparment of Anatomy, Ehime University Medical School).

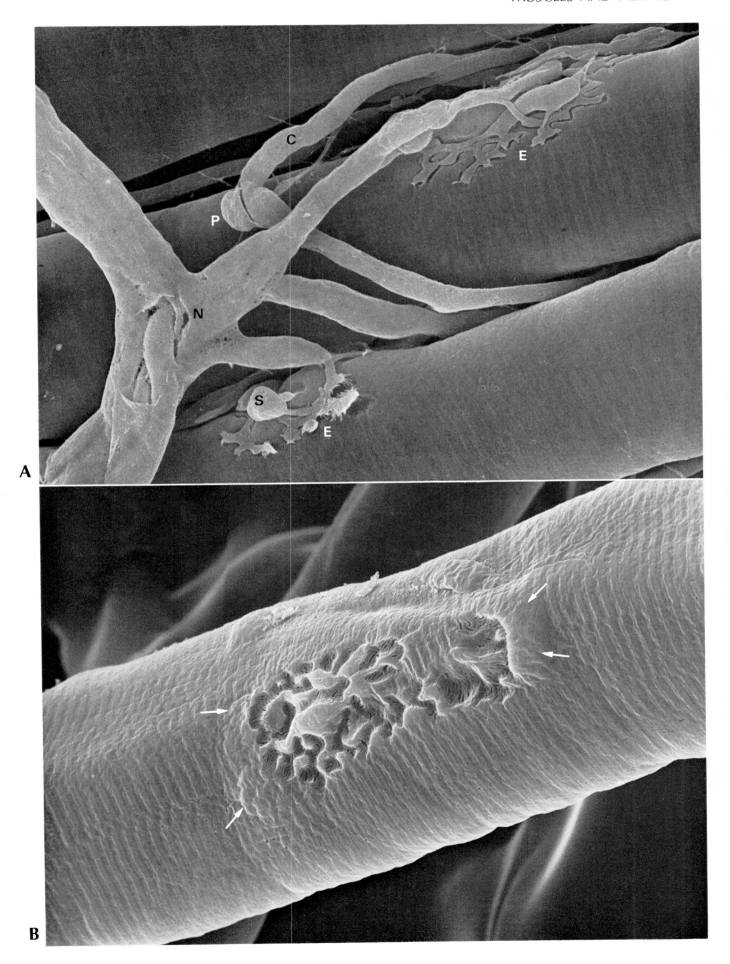

Plate XI-4   Silver-Impregnated Nerve.   Rabbit Heart.

Cardiac nerves were silver-impregnated and observed both by light microscope and by SEM. On the upper right a twisting nerve is seen in the light microscope showing blackened nerve fibers and nuclei of Schwann cells (A and B). The SEM image which was obtained at 25 kV of accelerating voltage clearly shows the light oval spots corresponding to the silver-blackened nuclei of Schwann cells.

X 1,800

Plate XI-5
A. Myelinated Nerve Fibers. Rat Sciatic Nerve.

A nerve was longitudinally split after dehydration of the specimen, and the surface of myelinated nerve fibers were exposed with a thin layer of collagen fibrils remaining on it. A *node of Ranvier* is indicated by an arrow.

   X 1,500

B. Cross Section of Nerve Fibers. Chicken.

A nerve was freeze-fractured transversely. An *axon* (A), *myelin* sheath (M) and occasionally hit *Schwann cell nucleus* (arrow) are the components of a myelinated nerve fiber. The myelin sheath partly shows lamellar structure which is accentuated by artifactual effects.

   Among the nerve fibers are collagen fibrils belonging to the endoneurium. The fibrils run mainly longitudinally.

   X 3,200

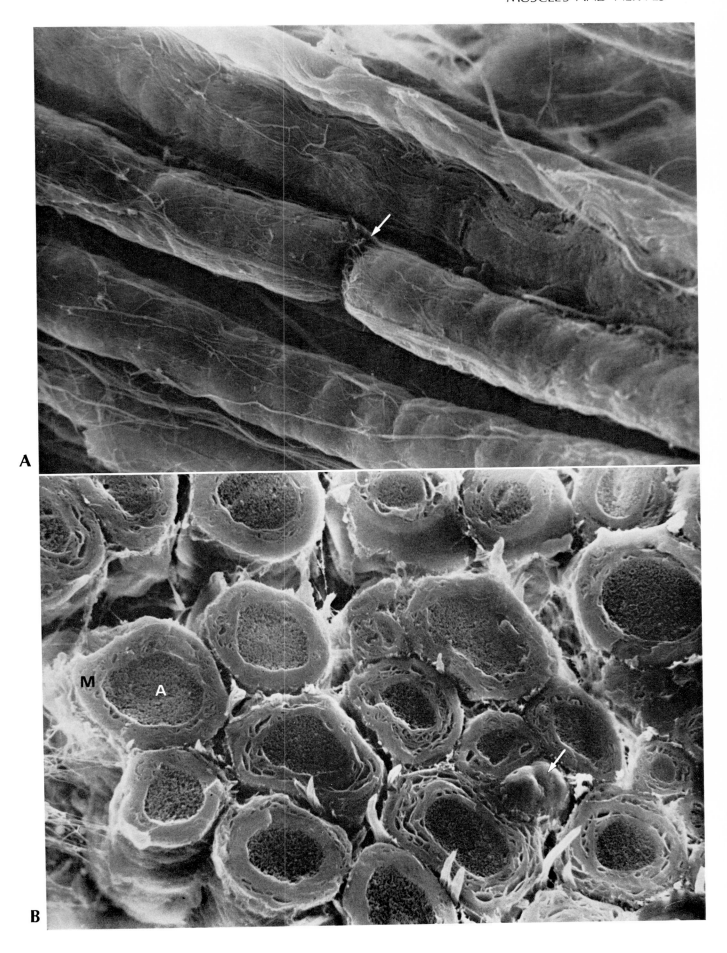

Brain

The surface structure of the brain ventricles including the choroid plexus has been a favorite object of SEM studies. Fairly numerous papers in this field are available (Mestres and Stumpf, 1978) and the attention of researchers is turning to the regional differences in the ventricular surface structure and to the circumventricular organs (Scott *et al.*, 1974; Krstić, 1981). Most parts of the brain ventricles are lined by ependymal cells with long cilla, and the SEM is useful for the analysis of the direction of *cillary beat* at given sites of the ventricles (Yamadori, 1977).

Plate XI-6   Ciliated and Non-Ciliated Ependyma.   Japanese Monkey.

A is the typical surface view of the lateral ventricle. *Cilia,* over 10 μm in length, are gathered at the center of the cell apical surface, which further is covered by delicate *microvilli.*

  B shows non-ciliated ependyma seen in the area postrema of the fourth ventricle. This and certain other portions including the infundibular recess of the third ventricle are known to be lacking in cilia and covered only by *microvilli.* The micrograph shows, besides typical microvilli, irregular globules of unknown nature. A probable single cilium (arrow) is seen.

  A: × 6,000,   B: × 25,000

A

B

Plate XI-7   Choroid Plexus.   Rat.

The choroid plexus occurs in the roof of the third and fourth ventricles and is known as the organ secreting the cerebrospinal liquor. The choroid plexus consists of blood vessels, mainly sinusoidal capillaries, and a cuboidal epithelium covering them, and the whole organ protrudes into the liquor space.

A is an overview of the choroid plexus which is highly complicated in surface showing strong foldings and protrusions. The light spots on the surface correspond to individual epithelial cells.

B shows the *epithelial cells* of the choroid plexus in closer view. These cells are characterized by *microvilli* which cover the cell surface very densely and which are polyp-like in shape showing a rounded swelling at the end. Furthermore, the cells possess a small bundle of *cilia* at the center of the swollen cell apex.

C shows a *macrophage* on the surface of the choroid plexus. The macrophage in the brain ventricles has long been known as the Kolmer cell and has been frequently studied by SEM since the report by Hosoya and Fujita (1973). These authors classified the cell into two types, one radiating numerous thin processes and the other projecting a few pseudopodial processes. The cell shown in this micrograph belongs to the first category (Type I of Hosoya and Fujita).

The macrophages in the ventricles including the choroid plexus surface are believed to patrol around and remove foreign bodies and dying cells.

A: X 165,   B: X 2,800,   C: X 5,700

(Plate XI-7: Courtesy of Profs. M. Murakami and T. Shimada, Department of Anatomy, Kurume University Faculty of Medicine)

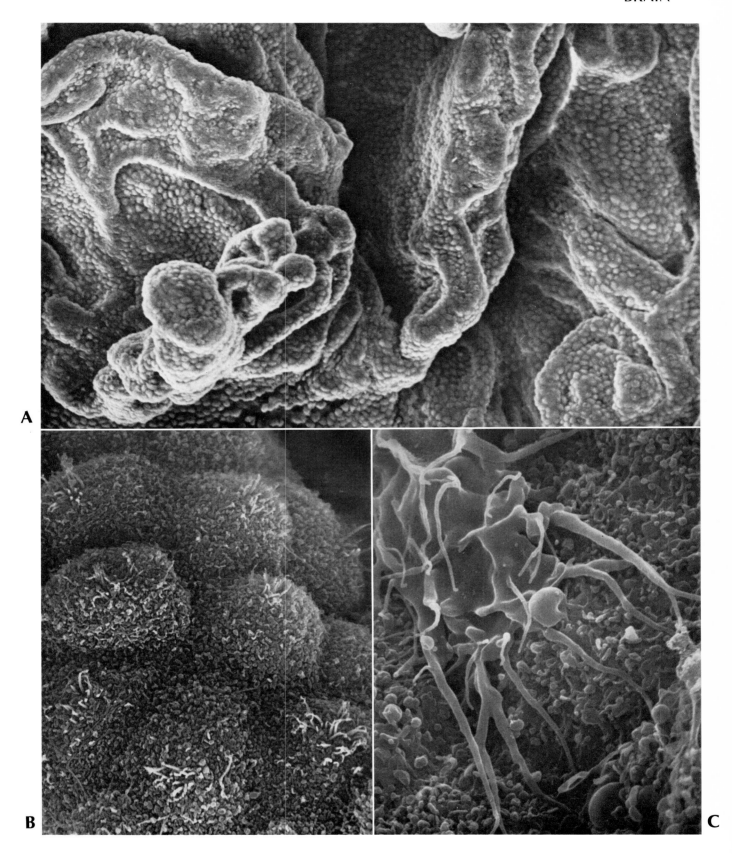

Plate XI-8   Brain Parenchyme.   Rat Cerebrum.

The freeze-cracked surface of the brain parenchyme is an unexplored but promising object of SEM study.

A shows a large *nerve cell* (N) surrounded by numerous *synaptic buttons* (arrows). A probable nerve cell process (n) is also found associated with synaptic buttons (arrows). Thick myelinated fibers (F), thin unmyelinated fibers and some blood vessels are identified.

B shows, in higher magnification, the architecture of the brain parenchyme. Attention should be paid to the *myelinated nerve fiber* (F) attached to arm-like processes of a cell (G) which is presumed to be a *glial cell,* possibly an astrocyte.

A: X 800,   B: X 15,500

(Plate XI-8: Micrographs prepared by Dr. M. Muto, Department of Anatomy, Niigata University School of Medicine).

References

Hosoya, Y. and T. Fujita: Scanning electron microscope observation of intraventricular macrophages (Kolmer cells) in the rat brain. *Arch. histol. jap.* **35**: 133—140 (1973).

Krstić, R.: Contribution of scanning electron microscopy to the study of brain ventricles, circumventricular organs and the pineal organ. *Biomed. Res.* 2, Suppl. : 129—137(1981).

Mestres, P. and W. E. Stumpf: Scanning electron microscope studies of the brain ventricular surfacrs. In: (ed. by) R. P. Becker and O. Johari: Scanning Electron Microscopy/ 1978/II. Scanning Electron Microscopy Inc. AMF O'Hare, 1978.

Scott, D. E., G. P. Kozlowski and M. N. Sheridan: Scanning electron microscopy in the ultrastructural analysis of the mammalian cerebral ventricular system. *Int. Rev. Cytol.* 37: 349—388 (1974).

Uehara, Y., J. Dezaki and T. Fujiwara: Vascular autonomic plexuses and skeletal neuromuscular junctions: A scanning electron microscopic study. *Biomed. Res.* 2, Suppl. : 139—143 (1981).

Yamadori, T.: The directions of ciliary beat on the wall of the fourth ventricle in the mouse. *Arch. histol. jap.* 40: 283—296 (1977).

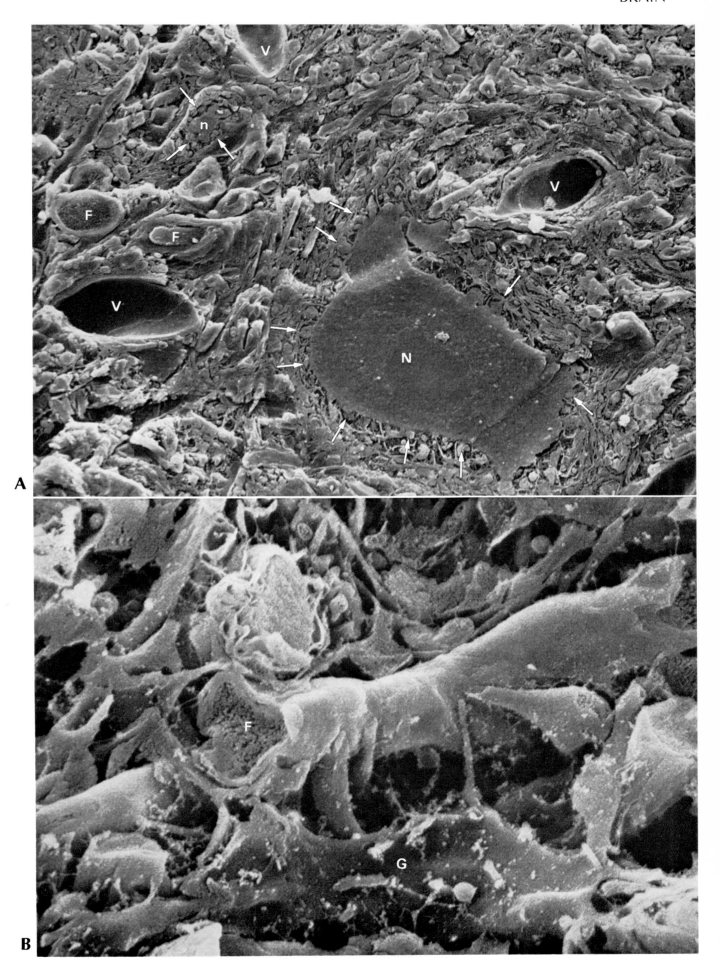

INDEX